Mathematical Engineering

Today, the development of high-tech systems is unthinkable without mathematical modeling and analysis of system behavior. As such, many fields in the modern engineering sciences (e.g. control engineering, communications engineering, mechanical engineering, and robotics) call for sophisticated mathematical methods in order to solve the tasks at hand.

The series Mathematical Engineering presents new or heretofore little-known methods to support engineers in finding suitable answers to their questions, presenting those methods in such manner as to make them ideally comprehensible and applicable in practice.

Therefore, the primary focus is—without neglecting mathematical accuracy—on comprehensibility and real-world applicability.

To submit a proposal or request further information, please use the PDF Proposal Form or contact directly: Dr. Thomas Ditzinger (thomas.ditzinger@springer.com)

Indexed by SCOPUS, zbMATH, SCImago.

More information about this series at https://link.springer.com/bookseries/8445

Andrea Bacciotti

Discrete Dynamics

Basic Theory and Examples

 Springer

Andrea Bacciotti
Dipartimento di Scienze Matematiche
"G.L. Lagrange"
Politecnico di Torino
Turin, Italy

ISSN 2192-4732   ISSN 2192-4740   (electronic)
Mathematical Engineering
ISBN 978-3-030-95094-1   ISBN 978-3-030-95092-7   (eBook)
https://doi.org/10.1007/978-3-030-95092-7

This Springer imprint is published by the registered company Springer Nature Switzerland AG
The registered company address is: Gewerbestrasse 11, 6330 Cham, Switzerland

Preface

Students who approach the theory of discrete dynamical systems may have the wrong impression that it is an easy topic. Indeed, at the beginning, only a few notions of elementary calculus (basically, the notion of limit) and a little familiarity with mathematical formalism and symbols are required. But going on, the width and the complexity of possible behaviors exhibited by such systems, sometimes unexpected and counterintuitive, rapidly emerge.

Discrete dynamical systems, as well as continuous time dynamical systems governed by ordinary differential equations, are often used to model physical systems of various nature. In fact, the same phenomenon can be often modeled either in the discrete time or the continuous time framework. However, these mathematically different representations not always are equivalent from the point of view of applications, because of the larger variety and generality of discrete time systems.

It is well known for instance that nonlinear discrete systems can exhibit characteristic traits of the so-called deterministic chaos also in dimension one while, in the continuous time case, chaotic behaviors can be observed only in dimension greater than or equal to three.

The purpose of this book is to provide a complete and detailed introduction to the theory of discrete dynamical systems, with special attention to stability of fixed points and periodic orbits. The aim is to supply a solid mathematical background and the essential basic knowledge for further developments such as, for instance, the aforementioned deterministic chaos theory, for which many other references are available (but sometimes, without an exhaustive presentation of preliminary notions). By the way, the reader will find in this book a discussion of some topics, sometimes neglected even in the research literature, such as a comparison between different predictions achievable by the discrete time model and the continuous time model of the same application. Another novel aspect of this book is an accurate analysis of the way a fixed point may lose stability, introducing and comparing several notions of instability: simple instability, repulsivity, complete instability. Even from this point of view, discrete systems present a more varied situation with respect to the continuous time ones.

The background required to the reader includes basic elementary calculus (for several variables), linear algebra and some notions about qualitative theory of ordinary differential equations. The book could be proposed as a reference for courses about dynamical systems in mathematics and theoretical engineering curricula.

To help the reader and to show the flexibility and potentiality of the discrete approach to dynamics, we inserted in the exposition many examples and simulations. Of course, some examples are academic, but for the most part they refer to applications. Apart from rather well-known examples taken from economy, epidemiology and population theory, we included also some new and rather surprising examples, like the construction of the music scale.

Other examples of systems of interest in physics and admitting a natural discrete time description can be found in the literature (for instance, the bouncing ball in [1] p. 103). Traditional engineering applications employ more frequently continuous time models involving ordinary or partial differential equations. However recently, because of the diffusion of computers and digital technologies, more attention has been deserved to the discrete time point of view (discretization, quantization and sampling methodologies in control and information theory, as explained for instance in [2]). We would like also recall the so-called Poincaré map: reducing the dynamics of a continuous time system to a discrete scheme, it provides a useful tool for studying the stability properties of limit cycles.

A short presentation of the contents of this book follows.

In Chap. 1, we introduce discrete dynamical systems and their solutions. We present some preliminary facts about certain special solutions such as fixed points and periodic orbits. We focus in particular on the general representation of the solutions of linear systems. Finally, we address the problem of the discretization (how to construct a discrete time model from a continuous time one) and the opposite problem (how to obtain a continuous time model from a discrete time one).

Chapter 2 is devoted to stability theory. After the introduction of the basic notions, we develop especially the Lyapunov function approach and the method of linearization. We study various versions of the Lotka–Volterra discrete model, included the standard SIR model for epidemiology. We also develop in a discrete setting an example taken by the engineering literature: the dynamics of a fluid in a chemical plant formed by a pair of communicating tanks.

Some notions of equivalence (global and local) for discrete systems are given in Chap. 3. Then we address the problem of the dependence on parameters of the qualitative dynamical behavior of such systems. For the one-dimensional case, we discuss the classical bifurcations (saddle-node, transcritical, pitchfork) and the related generality theorems. We discuss also the so-called flip bifurcation (period doubling) and, for the two dimensional case, the Neimark–Sacker bifurcation (secondary Hopf). We conclude the chapter by a short mention to chaotic dynamics.

The last chapter deals with linear systems defined by positive matrices. This is a classical subject, traditionally developed in probability courses (Markov chains) which has been recently revitalized with great success in social networks dynamics. Our exposition focuses in particular on the dynamical features of this class of systems.

This book is ideally dedicated to all my former students (of any level) and colleagues, who in several ways gave me stimulus and motivations. A special acknowledgment to Marco Gilli who raised my curiosity and interest in discrete dynamics during a teaching collaboration, Francesca Ceragioli for pointing out important references about Chap. 4 and Valeria Chiadò Piat for her encouragement. Finally, heartily thanks to my wife Giannina for her patience and support.

Turin, Italy   Andrea Bacciotti

References

1. Guckenheimer, J., Holmes, P.: Ordinary Differential Equations, Springer, New York (1983)
2. Jakubczyk, B., Sontag, E.D.: Controllability of nonlinear discrete-time systems: a Lie-algebraic approach. SIAM Journal Control and Optimization, **28**, 1–33 (1990)

Notations

- Throughout this book, $\mathbf{N}$ denotes the set of nonnegative integers i.e., $\mathbf{N} = \{0, 1, 2, 3, \ldots\}$, $\mathbf{Z}$ the set of integers, $\mathbf{R}$ the set of real numbers and $\mathbf{C}$ the set of complex numbers. By $|a|$ we denote the absolute value of a if $a \in \mathbf{R}$ or the modulus of a if $a \in \mathbf{C}$. If $a \in \mathbf{C}$, the conjugate, real part and imaginary part of a are respectively denoted by $\bar{a}$, $Re\ a$, $Im\ a$. The imaginary unit is represented as i.

  The set of positive real numbers i.e., $\{a \in \mathbf{R} : a > 0\}$, is denoted by $\mathbf{R}_+$.

- $\mathbf{R}^d$ is the usual Euclidean space of dimension d $(d \geq 1)$. If S is a subset of $\mathbf{R}^d$, the interior, closure and boundary of S are respectively denoted by $Int S$, $Clo S$, $Fr S$. For $x \in \mathbf{R}^d$, the symbol $|x|$ represents the Euclidean norm of x. The distance between points $x, y \in \mathbf{R}^d$ is denoted $d(x, y) = |x - y|$. The distance between points and sets $d(x, S)$ is defined as $\inf_{y \in S}\{|x - y|\}$. Moreover, the open ball of center x and radius $r > 0$ is $\mathcal{B}(r, x) = \{y : |y - x| < r\}$. Points $x \in \mathbf{R}^d$ are usually thought of as column vectors; however, for notational convenience, we often write $x = (x_1, \ldots, x_d)$ when we need to display the components of x. In Chap. 4 we will make use of the infinity norm of vectors, defined as $|x|_\infty = \max|x_i|$.

- Let A be a $d \times d$ square matrix. We denote respectively by $tr A$, $det A$ and $rank A$ the trace, the determinant and the rank of A. A^t denotes the transpose of A. The symbols $\ker A$, $im A$ represent the kernel and the image of A. The identity matrix of dimension d is denoted by I_d, or simply by I when the dimension is clear from the context. When we need to specify the entries of a square matrix, we write $A = (a_{ij})$.

- If A is a $d \times d$ square matrix, the spectrum of A is denoted by $\sigma(A)$ and the spectral radius by $\rho(A)$. To denote the algebraic multiplicity and the geometric multiplicity of an eigenvalue λ of A we write $m_a(\lambda)$ and, respectively, $m_g(\lambda)$ (or, simply, m_a and m_g). Recall that for each λ, $1 \leq m_g \leq m_a$. The eigenvalue λ is called *algebraically simple* if $m_a(\lambda) = 1$, and it is called *geometrically simple* if $m_g(\lambda) = m_a(\lambda)$.

The spectral norm of A is denoted by $||A||_2$. Sometimes, we will use also other norms, like for instance the infinity norm $||A||_\infty = \max_i \sum_{j=1}^d |a_{ij}|$ and the Frobenius norm $||A||_F = \sqrt{\sum a_{ij}^2}$.

- Let A be a $d \times d$ square matrix. A is said to be *positive definite* if for each $x \in \mathbf{R}^d$ ($x \neq 0$) one has $x^t A x > 0$, and it is said to be *positive semidefinite* if $x^t A x \geq 0$ for each $x \in \mathbf{R}^d$. In a similar way, *negative definite* and *negative semidefinite* matrices are defined. Finally, a matrix A is *indefinite* if there exist points $x, y \in \mathbf{R}^d$ such that $x^t A x > 0$ and $y^t A y < 0$.

- Let $f(x) : \mathbf{R}^{d_1} \to \mathbf{R}^{d_2}$ be a differentiable function, and write $x = (x_1, \ldots, x_{d_1})$, $f(x) = (f_1(x), \ldots, f_{d_2}(x))$. We denote by $\frac{\partial f_i}{\partial x_j}(x)$ the partial derivative of f_i with respect to x_j, and by $D(f)(x)$ the Jacobian matrix of f, evaluated at x. Of course, if $d_1 = d_2 = 1$, the more usual notation $\frac{df}{dx}$ may be convenient. If $x(t) : \mathbf{R} \to \mathbf{R}^d$ ($d > 1$), we may also write $\dot{x}(t)$ instead of $\frac{dx}{dt}(x)$.

Sometimes one needs to partition the independent variable as a pair of subvectors. If $f(x) : \mathbf{R}^{d_1} \times \mathbf{R}^{d_2} \to \mathbf{R}^d$ and $x = (y, z)$ with $y \in \mathbf{R}^{d_1}$ and $z \in \mathbf{R}^{d_2}$, the Jacobian matrix of the function $y \mapsto f(y, z) : \mathbf{R}^{d_1} \to \mathbf{R}^d$ for a fixed z is written $D_y(f)(y, z)$.

Contents

Chapter 1
Discrete Dynamical Systems

A discrete time (or, simply, discrete) dynamical system is a rule that, when applied recursively, generates a sequence of numbers or vectors. In general, we assume that such a rule can be represented by a function

$$f : \Omega \to \mathbf{R}^d, \tag{1.1}$$

where $\Omega \subseteq \mathbf{R}^d$ ($\Omega \neq \emptyset$, $d \geq 1$ any positive integer). If not differently stated, we always assume that Ω is equipped with the induced topology of $\mathbf{R}^d$.

In this introductory chapter, we present the basic formalism about discrete dynamical systems and the general properties of their solutions. We focus in particular on certain special types of solutions (constant and periodic). Some more details are given for the cases of one-dimensional systems and linear systems. Finally, we compare discrete time systems with continuous time systems modeling the same application. We address the problem of the consistency of the conclusions resulting from the study of the two types of modelization. This Chapter contains also several illustrative examples.

1.1 General Notions and Examples

1.1.1 Discrete Systems and Their Solutions

Throughout these notes, using the notation (1.1) we implicitly mean that $f(x)$ is defined for each $x \in \Omega$. The sequence $\{x_n\}$ ($n \in \mathbf{N}$, $x_n \in \Omega$) is said to be generated by f if

$$x_{n+1} = f(x_n) \tag{1.2}$$

© The Author(s), under exclusive license to Springer Nature Switzerland AG 2022
A. Bacciotti, *Discrete Dynamics*, Mathematical Engineering,
https://doi.org/10.1007/978-3-030-95092-7_1

for each $n \in \mathbf{N}$. The identity (1.2) is usually adopted as a conventional representation of a *discrete dynamical system*. Alternatively, we may also use the notation

$$x^+ = f(x) . \tag{1.3}$$

In the early literature, a system of the form (1.2) is sometimes called a *finite difference equation*. If $\{x_n\}$ is generated by f, we also say that it is a *forward solution* of (1.2). Of course, to start the iterative process, an initial value $\hat{x} \in \Omega$ has to be assigned: the sequence $\{x_n\}$ generated by f and such that $x_0 = \hat{x}$ is identified by writing

$$\begin{cases} x_{n+1} = f(x_n) \\ x_0 = \hat{x} . \end{cases} \tag{1.4}$$

It is called the *forward solution* of the system (1.2) corresponding to (or issuing from) the initial value $\hat{x}$ or, simply, a *forward solution* of the problem (1.4). In what follows, if not necessary in order to avoid ambiguity, the qualifier "forward" will be omitted.

1.1.2   Existence and Representation

When dealing with system (1.2), we also implicitly assume that

$$f(\Omega) \subseteq \Omega . \tag{1.5}$$

Under this assumption, for every $\hat{x} \in \Omega$ there exists a unique solution $\{x_n\}$ of system (1.2), defined for all $n = 0, 1, 2, 3, \ldots$. Having in mind the iterative nature of a discrete dynamical system, it is clear that this solution can be represented writing

$$x_0 = \hat{x} , \quad x_1 = f(x_0) , \quad x_2 = f(x_1) = f(f(x_0)) = (f \circ f)(x_0) , \quad \ldots$$

where the symbol $\circ$ is the map composition. To simplify the notation we will write

$$f^{[n]} = \underbrace{f \circ \ldots \circ f}_{n \text{ times}}$$

so that $x_n = f^{[n]}(x_0)$, for $n = 1, 2, 3, \ldots$. We also agree to write, by consistency, $x_0 = f^{[0]}(x_0)$.

In applications, discrete dynamical systems are convenient to describe the state evolution of physical quantities whose magnitude is measured at fixed instants of time. This may results, for instance, from sampling or discretization processes. Having in mind this type of applications, the variable x is often referred to as the *state* of the system, the value $\hat{x}$ as the *initial state* and the index n as the *time*. In accordance, $\Omega \subseteq \mathbf{R}^d$ is also called the *state space*.

Example 1.1 Simple examples of discrete dynamical systems taken from elementary mathematics are the *arithmetic progressions*. They are defined as

$$x_{n+1} = x_n + a \qquad (1.6)$$

where $a \in \mathbf{R}$ is fixed, and $\Omega = \mathbf{R}$ ($d = 1$). The solution of (1.6) with $x_0 = \hat{x}$ has the form

$$x_n = \hat{x} + na \, , \quad n \in \mathbf{N} \qquad (1.7)$$

For instance, with $a = 1$ and $\hat{x} = 0$, (1.6) generates the usual sequence of natural numbers $1, 2, 3, \ldots$. In fact, (1.6) can be thought of as a way to construct the set of natural numbers, according to Peano's axioms.

We may note that (1.6) and (1.7) define the same mathematical object and, in this sense, they are equivalent. However, (1.7) is more explicit: it allows us to compute directly the value of x_n for any desired n without need of computing all the preceding values $x_1, x_2, \ldots, x_{n-1}$. ∎

Example 1.2 A *geometric progression* corresponds to a discrete dynamical system with $\Omega = \mathbf{R}$ ($d = 1$) of the form

$$x_{n+1} = ax_n \qquad (1.8)$$

where, as before, $a \in \mathbf{R}$ is a constant. The explicit representation of the solution of (1.8) with $x_0 = \hat{x}$ is

$$x_n = a^n \hat{x} \qquad (1.9)$$

for each $n \in \mathbf{N}$. The behavior of (1.9) depends on the value of a.

- If $a > 1$ the solutions are increasing and go to $+\infty$ if $\hat{x} > 0$; if $\hat{x} < 0$, the solutions are decreasing and go to $-\infty$.
- If $a = 1$, for every $\hat{x} \in \mathbf{R}$ the solution is constant.
- If $0 < a < 1$ the solutions converges to zero (increasing if $\hat{x} < 0$, decreasing if $\hat{x} > 0$).
- If $a = 0$, for all $\hat{x} \in \mathbf{R}$ one has $x_0 = \hat{x}, x_1 = x_2 = \ldots = 0$.
- If $-1 < a < 0$ the solutions converges to zero but not monotonically (the solutions oscillate).
- If $a = -1$, for all $\hat{x} \neq 0$ we have $x_n = \hat{x}$ if n is even and $x_n = -\hat{x}$ if n is odd.
- If $a < -1$ and $\hat{x} \neq 0$, we have $\lim_{n \to +\infty} |x_n| = +\infty$, but the limit of the solution does not exist.

A historical example of geometric progression is Malthus' law, one of the simplest and earliest model in population theory. At the end of 18th century, T.R. Malthus conjectured that the number of individuals of a population, if its reproduction is not subject to constraints, undergoes an exponential grow (i.e., it grows according to a mathematical law of the form (1.9)). More precisely, assume that the rates of births and deaths registered during one unit of time are respectively equal to the $b_0\%$ and the $c_0\%$ of the existing population. Let $b = b_0/100, c = c_0/100$ and $a = 1 + b - c$. Note that $b \geq 0$, $c \in [0, 1]$ and $a \geq 0$. Denote by $\hat{x} \geq 0$ the initial number of individuals, and by $x_n \geq 0$ the number of individuals evaluated after n units of times. Then, we can cast the following predictions.

1. if $a > 1$ the population actually grows;
2. if $a = 1$ the number of individual remains constant;
3. if $a < 1$ the population is destined to the extinction.

The coefficient a is called the *incremental factor* of the population.   ∎

Remark 1.1 Systems of the form (1.2) are also called *first order* dynamical systems. In general a h-th order discrete dynamical system is a relation of the form

$$x_{n+h} = f(x_n, x_{n+1}, \ldots, x_{n+h-1}) .$$

In this case, to start the iteration we need to assign the initial values of $x_0, x_1, \ldots,$ x_{h-1}. A h-th order discrete dynamical system can be reduced to the form (1.2), provided that the dimension of the state space is augmented. For instance, if $h = 2$, we may introduce the new variable $z_n = x_{n+1}$, to get the system

$$\begin{cases} x_{n+1} = z_n \\ z_{n+1} = f(x_n, z_n) \end{cases}$$

in $\mathbf{R}^2$. This approach can be generalized to any integer h, with some notational complications.   ∎

Example 1.3 The sequence of Fibonacci numbers $1, 1, 2, 3, 5, 8, 13, \ldots$ is generated by the second order scalar system

$$x_{n+2} = x_{n+1} + x_n$$

with the initial data $x_0 = x_1 = 1$. It is equivalent to the system

$$\begin{pmatrix} x_{n+1} \\ z_{n+1} \end{pmatrix} = A \begin{pmatrix} x_n \\ z_n \end{pmatrix} \quad \text{where} \quad A = \begin{pmatrix} 0 & 1 \\ 1 & 1 \end{pmatrix} .$$

The interpretation of the Fibonacci sequence as solution of a dynamical system is well known and, as we shall see later, can help to understand and explain its interesting properties [1, 2].   ∎

Remark 1.2 Often, systems of the form (1.2) are also qualified as *time-invariant* (or *autonomous*). Instead, time-varying systems are defined by functions which explicitly depend on the variable n, that is

$$x_{n+1} = f(x_n, n) .$$

Even in this case, we can reformulate the problem as

$$\begin{cases} x_{n+1} = f(x_n, t_n) \\ t_{n+1} = t_n + 1 \end{cases}$$

where the new variable t is inizialized as $t_0 = 0$.   ∎

Example 1.4 A simple example of time-varying system is

$$x_{n+1} = (n + 1)x_n .$$

The solution such that $x_0 = 1$ is $x_n = n!$.   ∎

1.1.3 Invertible Systems

If $f(\Omega) = \Omega$ and the map f is invertible (i.e., bijective) on Ω, we may consider the system

$$y_{n+1} = f^{-1}(y_n) \tag{1.10}$$

as well. It is called the *reversed-time system* associated to (1.2). Any solution of (1.10) is also called a *backward* solution of (1.2). The reason of this terminology is that for each index m, if $\{x_n\}$ is the solution of (1.2) such that $x_0 = \hat{x}$, and $\{y_n\}$ is the solution of (1.10) such that $y_0 = x_m$, then $y_m = x_0$.

If f is invertible, since $(f^{[n]})^{-1} = (f^{-1})^{[n]}$ we may represent backward solutions by introducing the notation $(f^{[-n]}) = (f^{-1})^{[n]}$. Thus, forward and backward solutions of (1.2) corresponding to the same initial state $\hat{x}$ can be written in a unified way as $x_n = f^{[n]}(\hat{x})$, by allowing $n \in \mathbf{Z}$.

Example 1.5 A *musical scale* is a selection of sounds of appropriate frequencies. The selected sounds are called *notes*.[1] The basic consideration in order to construct a musical scale, is the notion of *pitch class*. Two notes belong to the same pitch class if the quotient between their frequencies is exactly an integer power of 2. Denoting by $\omega > 0$ the frequency of a note, the interval of frequencies $[\omega, 2\omega)$ is called an

---

[1] Literally, the term "notes" refers to the symbols adopted to identify the selected sounds, but it is commonly applied to the sound itself.

octave. The notes of a musical scale are included in an octave, and they are ordered according to their increasing frequencies.

We know that the human ear recognizes as equal the intervals between pairs of notes when they are "equal" in logarithmic sense i.e.,

$$[\omega_1, \omega_2] \text{ equal to } [\omega_1', \omega_2'] \iff \frac{\omega_2}{\omega_1} = \frac{\omega_2'}{\omega_1'}.$$

The ancient Pytagorean scale is constructed recursively, starting from a conventional choice of an initial frequency $\hat{\omega} > 0$, according to the following rule. If $\omega \in [\hat{\omega}, 2\hat{\omega})$ is a selected frequency, then also

$$\omega^+ = f(\omega) = \begin{cases} \dfrac{3}{2}\omega & \text{if } \hat{\omega} \leq \omega < \dfrac{4}{3}\hat{\omega} \\[3mm] \dfrac{3}{4}\omega & \text{if } \dfrac{4}{3}\hat{\omega} \leq \omega < 2\hat{\omega} \end{cases} \tag{1.11}$$

is a selected frequency (see Fig. 1.1). The rule (1.11) actually defines a dynamical system on the octave interval $[\hat{\omega}, 2\hat{\omega})$. Note that here the function f is discontinuous, but invertible. Traditionally, the Pytagorean scale stops when seven frequencies have been selected. For instance, if the chosen starting frequency $\hat{\omega}$ is an element of the pitch class **F** (in English notation), we obtain the following sequence

$$\omega_0 = \hat{\omega} = \mathbf{F}, \ \omega_1 = \mathbf{C}, \ \omega_2 = \mathbf{G}, \ \omega_3 = \mathbf{D}, \ \omega_4 = \mathbf{A}, \ \omega_5 = \mathbf{E}, \ \omega_6 = \mathbf{B}.$$

We remark that the generation of the musical notes by this procedure does not follow the natural order of the increasing frequencies. To obtain the usual sequence of notes (*diatonic scale*), a reordering is needed.

Of course, the iteration process can be pursued to generate the accidentals (denoted by superscripts ♭ or ♯) . The notes corresponding to the first 12 selected frequencies $\omega_0, \ldots, \omega_{11}$ form (after reordering) the *chromatic scale*. It would be natural to expect that $\omega_{12} = 2\hat{\omega}$: indeed, in this case we could further iterate the process continuing the chromatic scale on the following octave interval $[2\hat{\omega}, 4\hat{\omega})$ and so on. Unfortunately, ω_{12} turns out to be approximately equal to $2\hat{\omega}$, but not exactly equal. The quotient $\omega_{12}/2\hat{\omega}$ is approximately equal to 1.0136. This approximation error is called *Pytagorean comma*.

The notes of the chromatic scale provide a partition of $[\hat{\omega}, 2\hat{\omega})$ in 12 subintervals, called *semi-tones* (a pair of consecutive semi-tones is called a *tone*). Again, it happens that these subintervals are approximately, but not exactly, equal each other in logarithmic sense. Actually, there are two types of semi-tones; for instance, $\mathbf{D}^\flat/\mathbf{C} = 3^7/2^{11} = 1.067\ldots$, while $\mathbf{F}/\mathbf{E} = 2^8/3^5 = 1.053\ldots$.

The pitch class concept can be formalized by means of the equivalence relation on $\mathbf{R}_+$

$$b = a \sim_2 \iff \exists m \in \mathbf{Z} \text{ such that } b = 2^m a.$$

Fig. 1.1 Graph of the
function (1.11) with $\hat{\omega} = 1$

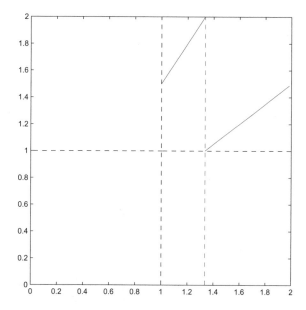

For any $\hat{\omega} > 0$, the interval $[\hat{\omega}, 2\hat{\omega})$ is a quotient set for the equivalence $\sim_2$. The system (1.11) can be rewritten as

$$\omega^+ = \left(\frac{3}{2}\omega\right) \sim_2 \tag{1.12}$$

with state space $\Omega = [\hat{\omega}, 2\hat{\omega})$. The solutions of (1.12) have the form

$$\omega_n = \frac{3^n}{2^m}\hat{\omega} \text{ for some } m \geq n \text{ such that } 2^m < 3^n < 2^{m+1} . \tag{1.13}$$

∎

1.1.4   Uniqueness

As already noticed, for every initial state $\hat{x} \in \Omega$ system (1.2) has a unique forward solution $\{x_n\}$ (i.e., defined for all future time $n = 1, 2, 3, \ldots$). Instead, the uniqueness is not guaranteed in the past, unless f is invertible. In other words, if a solution $\{x_n\}$ is known for every n greater than some $\bar{n}$, in general it is not possible to reconstruct the initial value where this solution is coming from.

Example 1.6 Let $f : \mathbf{R} \to \mathbf{R}$ be defined as

Fig. 1.2 Graph of the
function of Example 1.6

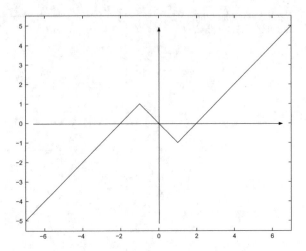

$$f(x) = \begin{cases} x + 2 & \text{for } x < -1 \\ -x & \text{for } -1 \le x \le 1 \\ x - 2 & \text{for } x > 1) \end{cases}$$

(see Fig. 1.2). Both the solutions issuing from $\hat{x} = -6$ and from $\hat{x} = 6$ vanish at the
third iteration.   ∎

1.1.5 Dependence on the Initial Data

Let $f : \Omega \to \mathbf{R}^d$ be continuous. Since the composition of continuous functions is
again a continuous function, the solutions of a discrete dynamic system, if f is
continuous, exhibit the property of *continuous dependence with respect to the initial
data*. This means that for each fixed n and for each $\varepsilon > 0$ there exists $\delta > 0$ such that

$$\hat{x}, \hat{y} \in \Omega , \ |\hat{x} - \hat{y}| < \delta \Longrightarrow |x_n - y_n| < \varepsilon$$

where of course y_n is the solution of (1.2) corresponding to the initial state $\hat{y}$. It is
important to remark that in general the value of δ depends not only on ε, but also on n.

1.1.6 Boundedness

A solution $\{x_n\}$ of (1.2) is said to be *forward bounded* (or, simply *bounded*) if there
exists a positive number L such that for each $n = 0, 1, 2, \ldots$, one has $|x_n| \le L$. This
means that there exists a bounded subset of $\mathbf{R}^d$ which contains the image of $\{x_n\}$.

Similarly, we may also define backward boundedness.

1.1.7 Orbits

The image in Ω of a solution $\{x_n\}$ of (1.2) is called a *positive orbit* (or, simply, *orbit*) of the system. The term *trajectory* is frequently used, but it is more ambiguous: it can be alternatively used to denote either a solution or an orbit. Note that an orbit, in general, is not a connected set.

It is important to distinguish the notion of "solution" from the notion of "orbit". A solution is a sequence i.e., a function from $\mathbf{N}$ to $\mathbf{R}^d$. Hence, it always has infinitely many values. However, as we shall see soon, it may happen that some of these values coincide; in fact, it may happen that the orbit is formed by a finite number of points.

When a system is invertible, we can define also negative orbits and complete orbits in the obvious way.

1.1.8 Invariance

We say that a subset $M \subset \Omega$ is *positively invariant* for (1.2) if

$$\forall x \in M \implies f(x) \in M. \tag{1.14}$$

As a consequence of (1.14),

$$\hat{x} \in M \implies x_n = f^{[n]}(\hat{x}) \in M$$

for all $n \in \mathbf{N}$. Orbits are examples of positively invariant sets. Condition (1.5) actually means that Ω is positively invariant. When f is invertible, we may define also negatively invariant sets and invariant sets. More precisely, we will say that M is *negatively invariant* if it is positively invariant for the associated reversed-time system, and that M is *invariant* if it is positively invariant for both the given system and the associated reversed-time system.

When $M \subset \Omega$ is positively invariant, it is possible to restrict the dynamical system to M without loosing the existence of the solutions.

1.1.9 Special Solutions

A point $x_* \in \Omega$ is said to be a *fixed point* of system (1.2) (or, in short, a fixed point of f) if $f(x_*) = x_*$. If x_* is a fixed point, then the solution generated by (1.2) taking x_* as initial state is constant.

A point $x_* \in \Omega$ such that $f^{[k]}(x_*) = x_*$ ($k \geq 2$), is said to be a *periodic point* of period k, and the solution corresponding to the initial state x_* is called a *periodic solution* of period k.

If $f^{[k]}(x_*) = x_*$ but $f^{[m]}(x_*) \neq x_*$ for each $m = 1, 2, \ldots, k - 1$, we say that k is the *minimal period*. By consistency, a fixed point can be considered a periodic point of period 1.

If x_* is a periodic point of period k, then also the points $f^{[m]}(x_*)$ for any $m = 1, \ldots, k - 1$ are periodic with the same period. Moreover, if x_* is a periodic point of period k, it is a periodic point of period $2k, 3k, \ldots$, as well. The following proposition is straightforward.

Proposition 1.1 *Let x_* be a periodic point of period k for system (1.2). Then x_* is a fixed point of the discrete dynamical system*

$$x^+ = f^{[k]}(x) \ .$$

The converse is true, as well.

A fixed point x_* is called *isolated* if there exists a neighborhood $\mathcal{U}$ of x_* such that for all $y \in \mathcal{U}$ except x_*, one has $y \neq f(y)$ i.e., y is not a fixed point. A similar definition applies to periodic points.

The orbit of a constant solution (fixed point) is a singleton. The orbit of a periodic solution is formed by finitely many points, and hence it is bounded. All of them can be thought of initial states of periodic solutions of (1.2) with the same period and describing the same orbit.

The orbit of a periodic solution of minimal period $k > 1$ is also called a *cycle* (of period k). If Γ is a cycle of (1.2) and $x_* \in \Gamma$, we also say that Γ is generated by (or issued from) x_*.

Example 1.7 For each $a \in \mathbf{R}$, a scalar system of the form (1.8) has a fixed point at the origin. This is actually the unique fixed point if $a \neq 1$. Instead, if $a = 1$ all the $\hat{x} \in \mathbf{R}$ are fixed points. If $a = -1$ all the nonzero $\hat{x}$ are periodic points of minimal period 2. Note that a system of the form (1.8) cannot have periodic points of minimal period $k > 2$. ∎

Example 1.8 The dynamical system

$$x^+ = \frac{1}{2}(1 - x)$$

($x \in \mathbf{R}$) has a unique fixed point at $x_* = 1/3$. A direct computation, based on the mathematical induction principle, shows that the solution corresponding to any initial state $\hat{x}$ is

$$x_n = \left(-\frac{1}{2}\right)^n \left(\hat{x} - \frac{1}{3}\right) + \frac{1}{3} \ . \tag{1.15}$$

Using this expression, it is easy to check that x_n converges to the fixed point x_* for each $\hat{x}$.

More generally, computing the solution of the dynamical system

$$x^+ = ax + (1-a)p$$

with $|a| < 1$ and arbitrary initial state, provides a way to approximate iteratively any real number p. ∎

Example 1.9 Many iterative algorithms studied in Numerical Analysis can be actually interpreted in terms of discrete dynamical systems. For instance, the so-called Hero's algorithm allows us to approximate the value of $\sqrt{a}$, being a any positive real number, by means of the solutions of the system

$$x^+ = \frac{1}{2}\left(x + \frac{a}{x}\right).$$

It is straightforward to check that $\sqrt{a}$ is the unique fixed point for this system. The tools which allows us to prove that nearby solutions actually approach the value $\sqrt{a}$ will be given in Chap. 2. ∎

Example 1.10 The system

$$x^+ = -x^3$$

($x \in \mathbf{R}$) has a unique fixed point at the origin. The point $\hat{x} = 1$ is a periodic point of minimal period 2. ∎

Example 1.11 Another well known dynamical system taken from population theory is the so called *logistic* or *Verhulst* model

$$x^+ = ax - \gamma x^2 \tag{1.16}$$

where a, γ are nonnegative parameters, and $x \in \mathbf{R}$. If $\gamma = 0$, (1.16) reduces to (1.8) and if in addition $a > 1$, the population x grows exponentially. The term $-\gamma x^2$ introduces an obstruction to the growth. For instance, it could model the fact that when x becomes greater and greater, the food provided by the environment becomes no more sufficient to guarantee the survival of all the individuals. For this reason, the coefficient γ is called the *intraspecific competition factor*.

From a mathematical point of view, system (1.16) is well defined for all $x \in \mathbf{R}$. However, a study of (1.16) with state space $\Omega = \mathbf{R}$ would have just an academic value. Indeed, it is clear that the model keeps a biological significance only if $x \geq 0$ and, unfortunately, the set $\{x \in \mathbf{R} : x \geq 0\}$ is not positively invariant for (1.16). This simple example enlightens the importance of investigating, as a first step, the existence of invariant sets.

Claim. If $\gamma > 0$ and $0 \leq a \leq 4$ the interval $[0, a/\gamma]$ is positively invariant for system (1.16).

The claim is readily proved remarking that the graph of the map $f(x) = ax - \gamma x^2$ is a parabola with a maximum at $x = a/2\gamma$ and that $f(a/2\gamma) = a^2/4\gamma$. If $a \leq 4$, then $a^2/4\gamma \leq a/\gamma$ (see Fig. 1.3).

Fig. 1.3 Function of
Example 1.11 with
$a = \gamma = 4$

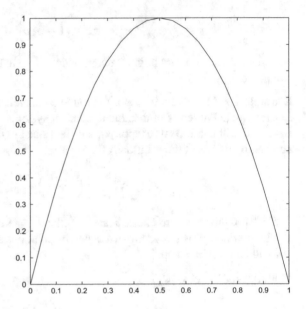

System (1.16) has two fixed points: the origin $p_1 = 0$, corresponding to the extinction of the population, and $p_2 = (a - 1)/\gamma$, corresponding to a level of ecological equilibrium. Both of them belong to the interval of invariance, provided that $a \geq 1$.

Now, take $a = \gamma = 4$. A simple computation shows that $f^{[2]}(x) = f(f(x)) = 16x(1 - x)(4x^2 - 4x + 1)$. The fixed points of the system

$$x^+ = f^{[2]}(x) \tag{1.17}$$

are the solutions of the algebraic equation

$$x(4x - 3)(16x^2 - 20x + 5) = 0 \,.$$

We recover the same roots $p_1 = 0$ and $p_2 = (a - 1)/\gamma = 3/4$ as before (being fixed points of (1.16), they must be fixed points of (1.17), as well). But we have in addition two new roots $q_1 = (5 - \sqrt{5})/8$ and $q_2 = (5 + \sqrt{5})/8$ (both in the interval of invariance), which represent, for (1.16), a pair of periodic points of period 2. If q_1 is taken as initial state for (1.16), the solution oscillates between q_1 and q_2. ∎

1.1.10 Changes of Variable

In the study of the behavior of the solutions of dynamical systems, it is often advisable to perform a preliminary change of variable in order, for instance, to move a fixed point in a more suitable position.

In general, let Ω, Ω' be open, connected subsets of $\mathbf{R}^d$. If $x = h(y) : \Omega' \to \Omega$ is an invertible map, we have

$$x^+ = h(y^+) = f(x) = f(h(y)) \,.$$

Setting $g(y) = h^{-1}(f(h(y))) : \Omega' \to \Omega'$, the original system (1.2) becomes

$$y^+ = g(y) \,. \tag{1.18}$$

More precisely, the sequence $\{y_n\}$ is the solution of the problem

$$\begin{cases} y^+ = g(y) \\ y_0 = h^{-1}(\hat{x}) \end{cases}$$

if and only if the sequence $\{x_n\}$, with $x_n = h(y_n)$, is the solutions of the problem (1.4).

It is clear that a transformation of this type maps fixed points of (1.2) into fixed points of (1.18). Indeed, if $x_* = f(x_*)$ and $x_* = h(y_*)$, then

$$g(y_*) = h^{-1}(f(h(y_*))) = h^{-1}(f(x_*)) = h^{-1}(x_*) = y_* \,.$$

Remark 1.3 A transformation between dynamical systems determined by an invertible map $x = h(y)$ is actually an equivalence relation. The transformation $x = h(y)$ is called a C^r-*equivalence* when h is at least of class C^r, with $r \geq 1$, together with its inverse. It is called a *topological equivalence* when h and h^{-1} are merely continuous i.e., when h is a homeomorphism. Finally, if h is a linear change of coordinates, we speak about *linear equivalence*. We will come back on this topic with more precision in Chap. 3. ∎

Example 1.12 A simple (but frequently applied) change of variable is the translation: $x = h(y) = y + v$, $v \in \mathbf{R}^d$ being a fixed vector. The transformed system takes the form

$$y^+ = g(y) = f(y + v) - v \,.$$

∎

Example 1.13 The reader probably wonders how it has been possible to conjecture the form (1.15) for the solutions of the system of Example 1.8. The trick is just a change of variable. Indeed, setting $y = x - 1/3$ the system becomes

$$y^+ = -\frac{1}{2} y$$

whose solution have the obvious expression $y = (-\frac{1}{2})^n \hat{y}$. ∎

Example 1.14 To carry on the analysis of a dynamical system, it is sometimes convenient to re-scale the state variable, namely to replace x by $\tilde{x} = cx$ for same suitable constant $c > 0$. We have:

$$\tilde{x}^+ = cx^+ = cf(x) = cf\left(\frac{\tilde{x}}{c}\right).$$

For instance, coming back on the logistic model discussed in Example 1.11, we have

$$\tilde{x}^+ = cx^+ = c(ax - \gamma x^2) = acx - \gamma cx^2 = a\tilde{x} - \frac{\gamma}{c}\tilde{x}^2$$

and taking in particular $c = \gamma/a$ we obtain the equation

$$\tilde{x}^+ = a\tilde{x}(1 - \tilde{x})$$

where only one parameter appears. Moreover, the positively invariant set where the biological significance is guaranteed becomes the interval $[0, 1]$. ∎

1.2 One Dimensional Systems

1.2.1 Systems on Intervals

When $d = 1$ the state variable is scalar and Ω is an interval (possibly unbounded). Many examples of this type have been already considered. The solutions of the discrete dynamical system (1.2) with $d = 1$ can be visualized by a particularly impressive graphic construction. Recall that the graph of the identity map $y = f(x) = x$ from $\mathbf{R}$ to $\mathbf{R}$ is the straight line bisecting the first and third quadrant of the xy-plane. The identity map can be viewed as a way to reflect the x-axis on the y-axis and vice-versa, by drawing vertical and horizontal lines. In particular, the fixed points of (1.2) are precisely the abscissæ (equivalently, the ordinates) of the points where the graph of f intersects the graph of the identity map. The periodic points of period 2 are the abscissæ (equivalently, the ordinates) of the points where the graph of $f^{[2]}$ intersects the graph of the identity map and so on.

Let $\hat{x}$ be the initial state. Let us draw a vertical line issuing from $\hat{x}$, until it hits the graph of f; then we continue with a horizontal line: the abscissa of the point P_1 where we meet the bisecting line is x_1. Starting from P_1, we now draw a new vertical line toward the graph of f followed as before by a horizontal line to meet the bisecting line. The abscissa of the point P_2 found in this way is x_2. By repeating this procedure, we can construct all the points x_n of the solutions (see Fig. 1.4) simply drawing a continuous line formed by vertical and horizontal segments.

Fig. 1.4 Construction of a
trajectory in the one
dimensional case

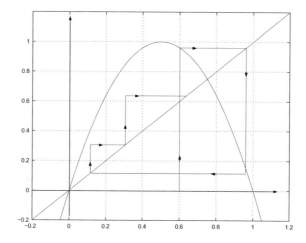

Example 1.15 This example aims to illustrate a type of solution with a particular
asymptotic property. Let us consider the system with $x \in \mathbf{R}$

$$
x^+ = f(x) = \begin{cases} \dfrac{1}{2}x & \text{if } x < 0 \\[2mm] 8x(1-x) & \text{if } x > 0\,. \end{cases}
$$

The origin is a fixed point. Let x_n be the solution corresponding to the initial
state $\hat{x} = 1/6$. Applying the graphic construction described in the previous para-
graph (or computing the values of x_n with the help of a computer), we realize that
$\lim_{n \to +\infty} x_n = 0$ (see Fig. 1.5). Although the function f is not invertible on its whole
domain, the solution issuing from $\hat{x} = 1/6$ admits a (not unique) reconstruction in
reversed time such that $\lim_{n \to -\infty} x_n = 0$. ■

Solutions which approach a fixed point both for $n \to -\infty$ and $n \to +\infty$ are
called *homoclinic*.

1.2.2 Systems on the Circle

In this section we are interested in the case where $\Omega = C \subset \mathbf{R}^2$, C being a circumference. The reader can conveniently think of C as the circumference centered at the
origin and of radius 1, since the exact length of the radius plays no essential role in
what follows. Let us denote by $\theta \in \mathbf{R}$ the arc length, measured in radiants starting
from the point of $\mathbf{R}^2$ of coordinates $(1, 0)$. Recall that the arcs θ_1, θ_2 are equivalent
modulo 2π if there exists $m \in \mathbf{Z}$ such that $\theta_2 - \theta_1 = 2m\pi$. This is written as

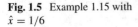

Fig. 1.5 Example 1.15 with $\hat{x} = 1/6$

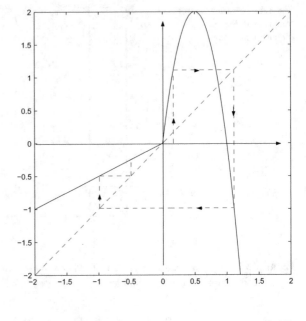

$$\theta_2 = \theta_1 \mathrm{mod}\,(2\pi)\,. \qquad (1.19)$$

By virtue of (1.19), we may agree that every point of C is identified by its principal value $\theta \in [0, 2\pi)$.

Example 1.16 Let $\alpha > 0$ be fixed, and consider the discrete dynamical system on C defined by

$$\theta_{n+1} = (\theta_n + \alpha)\mathrm{mod}\,(2\pi) \qquad (1.20)$$

whose solutions are easily found. They have the form $\theta_n = (\hat{\theta} + n\alpha)\,\mathrm{mod}\,(2\pi)$, for any initial value $\hat{\theta}$. Their qualitative properties depend on the nature of α (see Fig. 1.6).

- if $\alpha = 2p\pi$ for some $p \in \mathbf{N}$, every $\hat{\theta}$ is a fixed point;
- if $\alpha = 2\pi/q$ with $q \in \mathbf{N}$ and $q \geq 2$, after q iterations the system comes again at the initial position: all the points are periodic of minimal period q;
- if $\alpha = 2p\pi/q$ with $p, q \in \mathbf{N}$, $p, q \geq 2$ and co-prime, all the points are periodic of period q but, to come again at the initial position, the system turns around p times;
- if $\alpha/2\pi$ is not a rational number, there exist neither fixed nor periodic points: for every initial state, the corresponding orbit is dense in C. ∎

Example 1.17 In the light of the previous example, we revisit here the model of the ancient Pythagorean musical scale discussed in Sect. 1.1.3. Consider the system (1.12) and let us apply the transformation

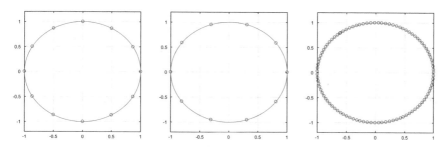

Fig. 1.6 Example 1.16 from left to right: $\alpha = 2\pi/12$, $\alpha = 6\pi/10$, $\alpha = 2\pi/(2\sqrt{2})$

$$\theta = 2\pi \log_2 \omega \qquad (1.21)$$

to the solution (1.13). We obtain

$$\theta_n = n(2\pi \log_2 3) + \hat{\theta} - 2m\pi .$$

This can be interpreted as the solution of a system on C of the form (1.20) with $\alpha = \log_2 3 = \log 2/\log 3$. This means that iterating the Pythagorean construction, dense orbits can be generated. In other words, after a sufficiently large number of iterations, we are able to obtain, potentially, sounds of almost all frequencies.

The Pythagorean scale construction was later replaced by the so-called well temperament. The well temperament can be modeled as a dynamical system, as well. More precisely, a well tempered musical scale is generated by a dynamical system of the the form

$$\omega_{n+1} = \left(\sqrt[12]{2}\, \omega_n \right)_{\sim 2} \qquad (1.22)$$

on some octave interval $[\hat{\omega}, 2\hat{\omega})$. Basically, it is therefore a geometric progression. The well temperament overcomes the drawbacks of the Pythagorean scale: the semitones are exactly equal (in logarithmic sense) and the Pythagorean comma disappears. Moreover, the notes are naturally determined in the order of increasing frequencies and can be continued in a unique way over the foregoing and following octaves.

Even in this case, we can reproduce on the circle C the solutions of (1.22)

$$\omega_n = \frac{(\sqrt[12]{2})^n}{2^m} \hat{\omega}$$

by the transformation (1.21). We obtain

$$\theta_n = \frac{2n\pi}{12} + \hat{\theta} - 2m\pi$$

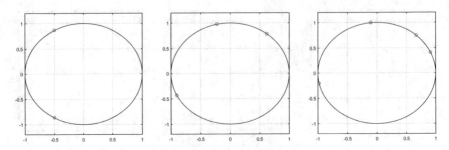

Fig. 1.7 Example 1.18 from left to right: orbits of period 2, 3, 4

which are the solutions of a system of the form (1.20) with $\alpha = 2\pi/12$. According to the discussion of the previous example, all the points are actually periodic of period 12.

We finally remark that reproducing a musical scale on C by means of the transformation (1.21), allows us to "hidden" the discontinuity on the right-hand-side of (1.11) (because of the different topologies of C and $[\hat{\omega}, 2\hat{\omega})$) and makes easier to change the octave interval.   ■

Example 1.18 Consider the dynamical system on C

$$\theta_{n+1} = 2\theta_n \bmod (2\pi) \, .$$

Now we have the unique fixed point $\theta = 0$. The explicit form of the solutions is $\theta_n = 2^n\theta_0 \bmod (2\pi)$.

We see that there are infinitely many periodic points. For instance $\theta_0 = \frac{2\pi}{3}$ of period 2, $\theta_0 = \frac{2\pi}{7}$ of period 3, $\theta_0 = \frac{2\pi}{15}$ of period 4, ...and, in general, $\theta_0 = \frac{2\pi}{2^n-1}$ of period n (see Fig. 1.7). There are also infinitely many points on C generating orbits which are dense on C (see Fig. 1.8).

This is probably the most elementary example of a system which exhibits deterministic chaos. As a consequence, this system exhibits also a strong sensitivity with respect to initial data (see Fig. 1.9).

The terms "deterministic chaos" and "sensitivity with respect to initial data" will be discussed with some more details at the end of Chap. 3. For the moment, we limit ourselves to emphasize that sensitivity with respect to initial data does not contradict the property of continuity with respect to initial data.   ■

Example 1.19 We give now another example of a system with homoclinic orbits. Consider the system

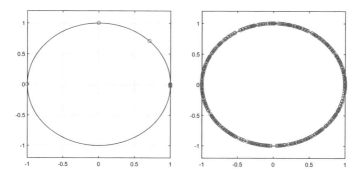

Fig. 1.8 Example 1.18 with $\theta_0 = \frac{\pi}{4} = 0.78539\ldots$ (50 iterations on the left; 500 iterations on the right)

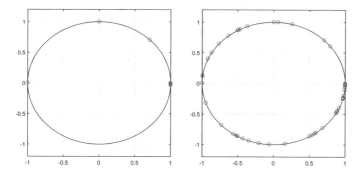

Fig. 1.9 Example 1.18: two solutions with slightly different initial states (50 iterations). $\theta_0 = \frac{\pi}{4} = 0.78539\ldots$ on the left, $\theta_0 = 0.785$ on the right

$$\theta^+ = f(\theta) = \theta + \frac{1}{2}\min\{\theta, 2\pi - \theta\} = \begin{cases} \dfrac{3}{2}\theta & \text{if } \theta \in [0, \pi] \\[2mm] \dfrac{\theta}{2} + \pi & \text{if } \theta \in [\pi, 2\pi) \end{cases} \qquad (1.23)$$

defined on the unit circle C. It is not difficult to check that if $\theta \in [0, 2\pi)$, then $f(\theta) \in [0, 2\pi)$, as well. So, the system is well defined. We notice that the function f is invertible, the inverse function being given by

$$\theta = f^{-1}(\theta^+) = \begin{cases} \dfrac{2}{3}\theta^+ & \text{if } \theta^+ \in [0, 3\pi/2] \\[2mm] 2(\theta^+ - \pi) & \text{if } \theta^+ \in [3\pi/2, 2\pi). \end{cases} \qquad (1.24)$$

Clearly, there is only one fixed point, $\theta_* = 0$. It is not difficult to see that the solution corresponding to any arbitrary initial state $\hat{\theta} \neq 0$ runs around the circum-

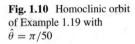

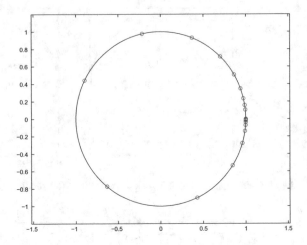

ference in the counterclockwise sense and eventually approaches the fixed point (see
Fig. 1.10). Symmetrically, from $\hat{\theta}$ we can go back, in the clockwise sense and reach
the fixed point asymptotically from the opposite side. Formally, we therefore have
$\lim_{n \to -\infty} \theta_n = 0 = \lim_{n \to +\infty} \theta_n$. ∎

1.3 Linear Systems

A discrete dynamical system for which $f(x) = Ax$, where A is a real square $d \times d$
matrix, that is

$$x_{n+1} = Ax_n \tag{1.25}$$

is said to be *linear*. Dealing with linear systems in abstract terms, we assume $\Omega = \mathbf{R}^d$.
From a mathematical point of view, linear systems constitute a rather simple (but
not trivial) class of systems; moreover, under certain circumstances and for certain
types of problems, linear systems can be used to approximate systems with a more
complicated structure, providing in this way useful information.

1.3.1 Invertibility

The case of a scalar (i.e., with $d = 1$) linear system

$$x^+ = ax \tag{1.26}$$

has been already introduced and analyzed in Examples 1.2 and 1.7. In (1.26), the map $f(x) = ax$ is always invertible, except for the case where $a = 0$. If $d > 1$, the map $f(x) = Ax$ is invertible if and only if the matrix A is nonsingular, that is if and only if zero is not an eigenvalue of A. Hence, when zero is an eigenvalue of A, we may expect that solutions with distinct initial states coalesce, so that the uniqueness of backward solution is lost.

Example 1.20 Let $f : \mathbf{R}^2 \to \mathbf{R}^2$ be the linear map defined by the matrix

$$A = \begin{pmatrix} -1 & 3 \\ -2 & 6 \end{pmatrix}$$

The eigenvalues are 0 and 5. For each initial state taken in the eigenspace associated to 0 (the line of slope $1/3$), the corresponding solution vanishes at the second iteration. ∎

If A is nilpotent, which means that all its eigenvalues are zero, this phenomenon occurs for each initial state.

Example 1.21 Let $f : \mathbf{R}^2 \to \mathbf{R}^2$ be the linear map defined by the matrix

$$A = \begin{pmatrix} 0 & 1 \\ 0 & 0 \end{pmatrix}$$

In this case, for every initial state the corresponding solution vanishes at the second iteration. ∎

1.3.2 Fixed and Periodic Points

A first aim of this section is a recognition of fixed points and periodic points of a linear system. The proofs of next two propositions are straightforward.

Proposition 1.2 *For any matrix A, the origin of $\mathbf{R}^d$ is a fixed point of (1.25). System (1.25) possesses nonzero fixed points if and only if $\lambda = 1$ is an eigenvalue of A. In this case, a nonzero vector $\hat{x} \in \mathbf{R}^d$ is a fixed point of (1.25) if and only if it is an eigenvector of A associated to the eigenvalue $\lambda = 1$.*

Proposition 1.3 *System (1.25) possesses nonzero periodic points of period k if and only if $\lambda = 1$ is an eigenvalue of A^k. In this case, a nonzero vector $\hat{x} \in \mathbf{R}^d$ is a periodic point of (1.25) of period k if and only if it is an eigenvector of A^k associated to $\lambda = 1$.*

From these propositions it follows that a linear system cannot have a finite number of isolated nonzero fixed or periodic points.

Example 1.22 The second order scalar system

$$x_{n+2} = \frac{x_n + x_{n+1}}{2}$$

generates sequences where each element is the arithmetic mean of the previous two. It is equivalent to a linear system in $\mathbf{R}^2$ with

$$A = \begin{pmatrix} 0 & 1 \\ \frac{1}{2} & \frac{1}{2} \end{pmatrix} .$$

The eigenvalues of this matrix are 1 and $-1/2$. The eigenvectors correspond-ing to $\lambda = 1$ are generated, for instance by $v = (1, 1)$. Hence, all the points of the form $(a, a) \in \mathbf{R}^2$ are fixed points. The eigenvectors corresponding to $\lambda = -1/2$ are generated, for instance by $v = (1, -1/2)$. ∎

Remark 1.4 It is well known that λ is en eigenvalue of A^k if and only if there exists an eigenvalue μ of A such that $\mu^k = \lambda$. Hence, if there is a nonzero periodic point of period k for (1.25) then there exists a eigenvalue μ of A such that $|\mu| = 1$. However, this condition is not sufficient. For instance, the matrix

$$A = \begin{pmatrix} \cos 1 & -\sin 1 \\ \sin 1 & \cos 1 \end{pmatrix}$$

has a pair of complex conjugate eigenvalues

$$\lambda = \cos 1 + i \sin 1 , \quad \bar{\lambda} = \cos 1 - i \sin 1$$

of modulus one. However, $\lambda^k = (\bar{\lambda})^k = \cos k + i \sin k \neq 1$ for each integer k. ∎

Example 1.23 Consider a linear system with

$$A = \begin{pmatrix} 0 & -1 \\ 1 & 0 \end{pmatrix} .$$

We have $A^4 = I$. Hence, all $x \in \mathbf{R}^2$ are periodic of minimal period 4. The origin is the unique fixed point. We may examine further this system by using polar coordinates

$$x = \rho \cos \theta , \quad y = \rho \sin \theta .$$

First we notice that

$$\rho^+ = \sqrt{(x^+)^2 + (y^+)^2} = \sqrt{x^2 + y^2} = \rho \tag{1.27}$$

which shows that every solution evolves over a circumference of constant radius. Then, from the system equations, we find

$$\begin{cases} \rho^+ \cos \theta^+ = -\rho \sin \theta \\ \rho^+ \sin \theta^+ = \rho \cos \theta \end{cases}$$

which yields, taking into account (1.27),

$$\text{tg}\, \theta^+ = -\cot g\, \theta = \text{tg}\left(\theta + \frac{\pi}{2}\right)$$

for $\theta^+ \neq \frac{\pi}{2} + k\pi, \theta \neq k\pi$ ($k \in \mathbf{Z}$). This implies that for each solution,

$$\theta^+ = \theta + \frac{\pi}{2}$$

where the identity should be intended $\mod(\pi)$. We therefore see that the solutions rotate, switching of an angle equal to $\pi/2$ at each step. ∎

1.3.3 Representation of Solutions

A second aim of this section is to study the set $\mathcal{G}$ formed by all the solutions of (1.25). It is called the *general solution* of (1.25) and it will be viewed as a subset of the vector space $\mathcal{S}$ formed by all the sequences $\{v_n\}$ of vectors of $\mathbf{R}^d$. We emphasize that $\mathcal{S}$ does not have a finite dimension.

We already noticed that when the state space is $\mathbf{R}$, a linear system reduces to a geometric progression. In fact, the representation of the solutions found in Example 1.2 readily extends to arbitrary dimension: we have

$$x_n = A^n \hat{x} . \tag{1.28}$$

This formula, when $\hat{x}$ varies in $\mathbf{R}^d$, provides a complete description of $\mathcal{G}$. However, it can be useful to give more explicit representations of $\mathcal{G}$ involving special solutions of relevant interest.

The first step in this direction is the following proposition which expresses a fundamental property of linear systems.

Proposition 1.4 *The set $\mathcal{G}$ constitutes a linear subspace of $\mathcal{S}$. In fact, the dimension of $\mathcal{G}$ is finite and it is equal to d.*

Proof Taking into account (1.28), it is clear that every linear combination of solutions of (1.25) is still a solution of (1.25). This proves the first statement. As far as the second statement is concerned, let $v_1, \ldots, v_d$ be a basis of $\mathbf{R}^d$ and let $\mathcal{F} = \{\{A^n v_1\}, \ldots, \{A^n v_d\}\}$. The elements of $\mathcal{F}$ are linearly independent in $\mathcal{S}$. Indeed, in the opposite case we should have for some scalar $c_1, \ldots, c_d$ (not all vanishing)

$$c_1 A^n v_1 + c_2 A^n v_2 + \ldots + c_d A^n v_d = 0$$

for each n, but this is impossible at least for $n = 0$. Now, let $\hat{x}$ be an arbitrary point of $\mathbf{R}^d$. We can write in a unique way $\hat{x} = a_1 v_1 + a_2 v_2 + \ldots + a_d v_d$. Then, by linearity, we have for each n

$$A^n \hat{x} = a_1 A^n v_1 + a_2 A^n v_2 + \ldots + a_d A^n v_d \,.$$

This proves that every solution of (1.25) can be expressed as a linear combination of elements of $\mathcal{F}$.   ∎

A basis of $\mathcal{G}$ is called a *fundamental set* of solutions. If a fundamental set of solutions is known, then any other solution can be easily recovered, according to the conclusion of Proposition 1.4. Now we illustrate how to obtain explicitly, by suitable choices of the initial state $\hat{x} \in \mathbf{R}^d$, a special basis of $\mathcal{G}$, related to characteristic elements of A like eigenvalues and eigenvectors.

First we note that if λ is a real eigenvalue of A and $v \neq 0$ is any corresponding eigenvector, then we have a solution $x_n = \lambda^n v$. In particular, as already remarked, if A has an eigenvalue $\lambda = 1$ with corresponding eigenvector v, then the solution is constant.

Example 1.24 Let us consider the linear system in $\mathbf{R}^2$

$$x_{n+1} = A x_n \tag{1.29}$$

with $A = \begin{pmatrix} 0 & 1 \\ 1 & 1 \end{pmatrix}$. Recall that this matrix has been already encountered in Example 1.3. The eigenvalues of A are

$$\lambda_1 = \frac{1 + \sqrt{5}}{2}, \quad \lambda_2 = \frac{1 - \sqrt{5}}{2}$$

and the respective eigenvectors are

$$v_1 = \begin{pmatrix} 1 \\ \frac{1+\sqrt{5}}{2} \end{pmatrix}, \quad v_2 = \begin{pmatrix} 1 \\ \frac{1-\sqrt{5}}{2} \end{pmatrix} \,.$$

All the solutions can be expressed as linear combination of the two fundamental solutions $\lambda_1^n v_1$ and $\lambda_1^n v_1$ that is,

$$x_n = c_1 \lambda_1^n v_1 + c_2 \lambda_2^n v_2 \tag{1.30}$$

for arbitrary constants c_1, c_2. Note that since $\lambda_1 > 1$ and $-1 < \lambda_2 < 0$,

$$\lim_{n \to +\infty} |\lambda_1^n v_1| = \infty \quad \text{and} \quad \lim_{n \to +\infty} |\lambda_2^n v_2| = 0 \,.$$

We can now easily prove the following statement.

Claim. For $n \to \infty$, and except for the case $c_1 = 0$, the norm of any solution x_n of the system (1.29) goes to the infinity while the direction of x_n asymptotically approaches the line of slope $\frac{1+\sqrt{5}}{2}$.

It is interesting to reinterpret the previous Claim in terms of the solutions of the scalar second order equation which system (1.29) cames from and which we rewrite here for convenience as

$$\xi_{n+2} = \xi_n + \xi_{n+1} . \tag{1.31}$$

The solutions of (1.31) are obtained by extracting the first component of the solutions of (1.29). For instance, choosing $c_1 = \lambda_1/\sqrt{5}$ and $c_2 = -\lambda_2/\sqrt{5}$, ξ_n provides the well known Fibonacci's numbers sequence. The previous Claim amounts to say that for each solution x_n of (1.29) (except for $c_1 = 0$) one has

$$\lim_{n \to +\infty} \frac{\xi_{n+1}}{\xi_n} = \frac{1+\sqrt{5}}{2} .$$

This fact is well known for the particular case of Fibonacci numbers; it was noticed for the first time by J. Kepler. Note finally that λ_1 coincides with the so called *golden ratio*. ∎

Example 1.25 Consider again the system of Example 1.22. We already computed eigenvalues and eigenvectors of the matrix A. We can therefore write the general solution

$$x_n = c_1 \begin{pmatrix} 1 \\ 1 \end{pmatrix} + c_2 \left(-\frac{1}{2} \right)^n \begin{pmatrix} 1 \\ -\frac{1}{2} \end{pmatrix}$$

for arbitrary constants c_1, c_2. ∎

If A has a complex (non real) eigenvalue $\lambda = \alpha + i\beta$ with eigenvector $v = u + iw$, then we may still write a solution as $x_n = \lambda^n v$. On the other hand, being A real, we also have the eigenvalue $\bar{\lambda} = \alpha - i\beta$ corresponding to the eigenvector $\bar{v} = u - iw$, which gives rise to a distinct solution $\bar{x}_n = \bar{\lambda}^n \bar{v}$. In fact, since β and w are not zero, these two solutions are linearly independent. Taking a suitable linear combination, the complex valued solutions x_n and $\bar{x}_n$ can be replaced by the pair of real valued solutions

$$y_n = \rho^n [\cos(n\varphi)u - \sin(n\varphi)w] , \quad z_n = \rho^n [\sin(n\varphi)u + \cos(n\varphi)w]$$

where $\rho = \sqrt{\alpha^2 + \beta^2}$ and $\varphi = \text{arctg} (\beta/\alpha)$. Note that y_n and z_n are periodic if and only if $\rho = 1$ and $\varphi/2\pi$ is rational.

Example 1.26 The second order scalar system

$$x_{n+2} = x_{n+1} - x_n$$

generates sequences where each element is the difference of the previous two. It is equivalent to a linear system in $\mathbf{R}^2$ with

$$A = \begin{pmatrix} 0 & 1 \\ -1 & 1 \end{pmatrix}.$$

The eigenvalues are $\lambda = \frac{1+i\sqrt{3}}{2}$ and $\bar{\lambda} = \frac{1-i\sqrt{3}}{2}$ with eigenvectors $u = (1, \frac{1+i\sqrt{3}}{2})$ and, respectively, $\bar{u} = (1, \frac{1-i\sqrt{3}}{2})$. The corresponding real valued independent solutions are

$$y_n = \begin{pmatrix} \cos\frac{n\pi}{3} \\ \frac{1}{2}\cos\frac{n\pi}{3} - \frac{\sqrt{3}}{2}\sin\frac{n\pi}{3} \end{pmatrix} \qquad z_n = \begin{pmatrix} \sin\frac{n\pi}{3} \\ \frac{\sqrt{3}}{2}\cos\frac{n\pi}{3} + \frac{1}{2}\sin\frac{n\pi}{3} \end{pmatrix}.$$

Note that $|\lambda| = |\bar{\lambda}| = 1$. All the nonzero solutions are periodic of period 6. ∎

The previous considerations allow us to construct a fundamental set of solutions, provided that the matrix A admits a proper basis (namely, a basis of $\mathbf{R}^d$ formed by eigenvectors). If A possesses some eigenvalue which is not geometrically simple, then a proper basis does not exists. An additional number of special solutions, sufficient to complete the construction of a fundamental set, can be obtained resorting to generalized eigenvectors. For instance, assume that this situation occurs for an eigenvalue λ. Let v be an eigenvector and $\tilde{v}$ a generalized eigenvector generated by v, namely

$$Av = \lambda v, \qquad A\tilde{v} = \lambda\tilde{v} + v.$$

Then, the sequence $\tilde{x}_n = \lambda^{n+1}\tilde{v} + (n+1)\lambda^n v$ is a solution, linearly independent from $x_n = \lambda^n v$.

Example 1.27 The matrix

$$A = \begin{pmatrix} 2 & 1 \\ 0 & 2 \end{pmatrix}$$

has a unique eigenvalue $\lambda = 2$ whose algebraic multiplicity is 2. The eigenspace has dimension 1, and it is generated, for instance, by the eigenvector $u = (1, 0)$. As a generalized eigenvector we can take $v = (0, 1)$. A fundamental set of solutions is formed by

$$x_n = \begin{pmatrix} 2^n \\ 0 \end{pmatrix}, \qquad \tilde{x}_n = \begin{pmatrix} (n+1)2^n \\ 2^{n+1} \end{pmatrix}$$

We do no treat further this topics: for a more complete description see [1, 3].

1.3.4 Linear Equivalence

Two linear systems $x^+ = Ax$ and $y^+ = By$, where A and B are real $d \times d$ square matrices are said to be *linearly equivalent* if they can be transformed each other by means of a linear change of variable $h(y) = Py$, where P is a real nonsingular $d \times d$ matrix. In this case, we have $B = P^{-1}AP$. In other words, such systems are linearly equivalent if and only if they are defined by similar matrices.

When convenient, linear equivalence allows us to assume without loss of generality, that A is written in some convenient form, for instance the Jordan canonical form.

1.3.5 Changing the Unit of Time

Usually, in application, the index n represents a measure of some physical quantity, for instance the time. The meaning of this index and the information provided by the study of the system are therefore depending on the choice of a unit of measure. Sometimes, it may be convenient to change it, for example in the case we need to compare the behavior of different models determined on the base of different units. Of course, we expect that the change of unit can be performed without affecting the features of the evolution.

Example 1.28 Consider the scalar linear system

$$x^+ = ax \tag{1.32}$$

and assume for simplicity that $a > 0$. We want to determine another system which describes the same time evolution as (1.32) (imposing of course, the same initial state $\hat{x}$) but with respect to a different choice of the unit of time. Assume for instance that we want to redefine a new unit as a fraction $1/k$ of the previous unit (here, $k > 1$ is an integer number). The problem consists in finding a number q such that for the solution y_m of the problem

$$y^+ = qy, \quad y_0 = \hat{x} \tag{1.33}$$

we have $y_{kn} = x_n$ for each n. To meet the required condition, we should have $q^{kn} = a^n$ for each n, which implies $q^k = a$, and finally we find $q = a^{1/k}$. $\blacksquare$

Example 1.29 We want to deposit money to a bank. The bank offers the following choices.

1. An annual compound interest of 4%, credited yearly.
2. A half-yearly compound interest of 1.9%, credited six-monthly.

In both cases, the deadline for withdraw the money is at the end of each year. To try to understand what is the more convenient offer, we may resort to discrete dynamical systems. The first offer gives rise to the system

$$y_{k+1} = y_k + \frac{4}{100} \, y_k = 1.04 \, y_k \, .$$

The second offer is represented by

$$x_{n+1} = x_n + \frac{1.9}{100} \, x_n = 1.019 \, x_n \, .$$

We are so led to compare the values of the sequences x_{2k} and y_k for the same initial datum. It is easy to check that, for each k,

$$(1.019)^{2k} < (1.04)^k \, .$$

∎

1.4   From Continuous to Discrete and Back

Many phenomena of the real world exhibiting a time evolution can be mathematically modeled either as discrete dynamical systems or as continuous dynamical systems. In this section, we aim to discuss the relationship between these possible representations, and how they can be transformed each other.

1.4.1   Continuous Time Systems

A continuous time (or, simply, continuous) dynamical system is defined by a system of ordinary differential equations of the form

$$\dot{x} = g(x) \tag{1.34}$$

where the dot indicates the derivative of the unknown function $x(t)$. Sometimes, system (1.34) will be referred to also as *differential system*. Here, we assume that $\Omega \subseteq \mathbf{R}^d$ is open and connected and that $g : \Omega \to \mathbf{R}^d$ is a function of class C^1. The map $(t, x) \mapsto \phi(t, x)$ such that for each $\hat{x} \in \Omega$, $x(t) = \phi(t, \hat{x})$ represents the solution of (1.34) for which $x(0) = \hat{x}$, is called the *flow map*. We always assume that the flow map is defined for each $t \in \mathbf{R}$, and that $x \in \Omega \implies \phi(t, x) \in \Omega$ for each $t \in \mathbf{R}$.

Using the same terminology as in the discrete case, the image of a solution is called an *orbit*. However, orbits of continuous dynamical systems, being images of continuous functions, are connected. To this respect, discrete dynamical systems and differential systems crucially differ.

Orbits deserving special attention in the case of continuous dynamical systems are the *equilibria*, that is points $\hat{x}$ where $g(\hat{x}) = 0$ (constant solutions), and periodic orbits, that is orbits represented by closed curves. Nontrivial periodic orbits of (1.34) can exist only if $d > 1$. This is another important difference with respect to the discrete case.

Example 1.30 The natural continuous time analog of a linear discrete one-dimensional system (1.32) is the scalar linear differential equation

$$\dot{x} = ax \tag{1.35}$$

where a is a given real constant. The general solution of (1.35) is $x(t) = e^{at}\hat{x}$. If $a = 0$, all the solutions are constant. If $a \neq 0$ the unique constant solution is $x(t) = 0$, and the system admits only three (complete) orbits: $\{x \in \mathbf{R} : x < 0\}$, $\{0\}$, $\{x \in \mathbf{R} : x > 0\}$. More precisely, taking distinct initial conditions with the same sign, we obtain distinct solutions, but all run the same orbit. We observe here another difference with respect to the discrete system (1.32). Indeed, when $a \neq 0$ the orbits of (1.32) issuing from $\hat{x}$ and $\hat{x}'$ coincide if and only if there exists $n \in \mathbf{Z}$ such that $a^n\hat{x} = \hat{x}'$: otherwise, they are distinct. Hence, for any value of $a \neq 0$ i.e., when it is invertible, system (1.32) always has infinitely many complete orbits.

■

1.4.2 Discretization

Let us consider the system (1.34). Let $(t, x) \mapsto \phi(t, x)$ be the associated flow map defined on $\mathbf{R} \times \Omega$ and let $\tau > 0$ be fixed. We say that the discrete system (1.2) with

$$f(x) = \phi(\tau, x) \tag{1.36}$$

is the τ-*sampled* (or *discretized*) version of (1.34). Of course, we have $x_n = \phi(n\tau, x)$ for each x. In this way, the solutions of the discretized system match exactly (at integer multiples of τ) the solutions of the continuous time system. Note in particular that if x_* is an equilibrium point of (1.34), then it is also a fixed point of its discretization for each τ.

We stress that the map f defined in (1.36) turns out to be invertible. Hence, the solutions of the discretized dynamical system are uniquely defined both in the forward and backward direction of time, as well as the solutions of the continuous system (1.34). The discrete systems which can be obtained starting with a continuous system by means of a discretization procedure constitute therefore a proper subset of the set of all the systems of the form (1.2). This confirms the impression that discrete dynamical systems are more general than their continuous time counterparts, and may present a larger variety of qualitative behaviors. Sampled versions are often very useful in the study of certain theoretical properties of continuous dynamical systems. For instance, it is well know that a periodic solution $x(t)$ of a differential system can be investigated by the so-called Poincaré map. The Poincaré map consists in a discretization of type (1.36) where τ is taken equal to the period of $x(t)$.

We notice that to perform a discretization procedure, one needs to know the explicit solutions of (1.34).

Example 1.31 The discretization of the scalar linear system (1.35) is the system $x^+ = e^{\tau a}x$.

■

1.4.3 Euler Approximation

The numerical approximation of the solutions of ordinary differential equations is a typical task of Numerical Analysis. The simplest but fundamental method goes back to Euler. It can be reinterpreted in terms of discrete dynamical systems.

Let $x(t)$ be the solution of (1.34) corresponding to an initial state $\hat{x}$. Let us fix an increment $h > 0$. Then, for any instant t, we replace in (1.34) the derivative by the difference quotient. We get

$$\frac{x(t+h) - x(t)}{h} = g(x(t))$$

or

$$x(t+h) = x(t) + hg(x(t)) . \tag{1.37}$$

In accordance to (1.37), choosing $t = nh$ with $n = 0, 1, 2, \ldots$, we can construct recursively the sequence $x_0 = x(0) = \hat{x}$, $x_1 = x_0 + hg(x_0)$, $x_2 = x_1 + hg(x_1)$, ... In general,

$$x_{n+1} = x_n + hg(x_n) . \tag{1.38}$$

We can think of x_1 as an approximation of $x(h)$, x_2 as an approximation of $x(2h)$, and so on. Connecting by straight lines the points of coordinates $(0, x_0)$, (h, x_1), $(2h, x_2)$, ..., we obtain a piecewise linear function $\psi_h(t)$. It is well known that for each $T > 0$ (and under some assumptions which guarantee the continuability of the solutions of (1.34))

$$\lim_{N \to +\infty} \psi_{T/N}(t) = x(t)$$

uniformly on the compact interval $[0, T]$.

Example 1.32 Let A be a real $d \times d$ matrix, and consider a linear differential system in $\mathbf{R}^d$

$$\dot{x} = Ax$$

whose general solution is $x(t) = e^{tA}\hat{x}$. The Euler approximation method in this case leads to the discrete system

$$x_{n+1} = (I + hA)\, x_n\, .$$

It generates the solution

$$x_n = (I + hA)^n\, \hat{x}\, . \tag{1.39}$$

If, for any fixed $T > 0$ and any integer $N > 0$, we take $h = T/N$, then (1.39) becomes

$$x_N = \left(I + \frac{T}{N}A\right)^N \hat{x} = \left[\left(I + \frac{T}{N}A\right)^{\frac{N}{T}}\right]^T \hat{x}$$

which converges to $x(t) = e^{TA}\hat{x}$ when $N \to \infty$, as expected. ∎

1.4.4 The Proportional Increment Hypothesis

In applied sciences, the construction of a mathematical model is often performed, at the first step, having in mind a discrete time scheme. Only at a second step, a continuous time model is deduced from the discrete one.

First step. Assume that a unit of measure for the time has been fixed and that the process is monitored starting at $t = 0$. Let us denote by x the quantity we are interested in. The measured value of x is recorded at each integer multiple of the time unit. We assume further that for each instant t, the increment of x during an interval of time $[t, t + 1]$ depends only on the value of x at t, namely

$$x(t + 1) = x(t) + \varphi(x(t))\, . \tag{1.40}$$

Writing $x_n = x(n)$, we arrive to the discrete dynamical system

$$x_{n+1} = x_n + \varphi(x_n)\, . \tag{1.41}$$

Second step. For an arbitrary instant t and an arbitrary interval $[t, t + h]$ $(h > 0)$, equation (1.40) is replaced by

$$x(t + h) = x(t) + h\varphi(x(t)) \tag{1.42}$$

which yields

$$\frac{x(t+h) - x(t)}{h} = \varphi(x(t))$$

and taking the limit for $h \to 0$ we finally get

$$\dot{x}(t) = \varphi(x(t)) .$$

The reasoning above is simple and frequently applied in the literature. By the way, it looks like an inverse procedure of the Euler approximation method. However, it should be noticed that the step from (1.40) to (1.42) rests on an implicit assumption namely, that the increment of x relative to a unit of time results from the accumulation of smaller increments relative to subintervals and proportionally distributed. In what follows, we refer to such assumption as the *proportional increment hypothesis*. We stress that the proportional increment hypothesis is questionable even in the linear case.

Example 1.33 The Malthus' model of population growth has been discussed in Example 1.2. It is represented by a scalar linear system

$$x_{n+1} = ax_n = (1+r)x_n \tag{1.43}$$

where a is a constant, and it is therefore of the form (1.41) with $d = 1$, $\varphi(x) = rx$. Assume for simplicity $r > -1$. Under the proportional increment hypothesis, the reasoning above leads to the differential equation

$$\dot{x} = rx \tag{1.44}$$

whose general solution has the form $x = e^{rt}c$, c being an arbitrary constant.

Note that the transformation from (1.40) to (1.42) can be thought of as a change of the unit of time. Having in mind the discussion of Example 1.28, a more correct equation to write should be

$$x(t+h) = (1+r)^h x(t) \tag{1.45}$$

which leads to

$$\frac{x(t+h) - x(t)}{h} = \frac{(1+r)^h - 1}{h} x(t) .$$

Finally, taking the limit for $h \to 0$, we obtain the equation

$$\dot{x} = \log(1+r)x . \tag{1.46}$$

The general solution of (1.46) is $x = (1+r)^t c$ which matches correctly the solutions of (1.43) for $t = n$. ∎

Unfortunately, it is not clear how to apply in general, for nonlinear systems, the approach illustrated in the previous example.

Example 1.34 Under the proportional increment hypothesis, from the discrete logistic equation (1.16) the following continuous time model in **R** is easily deduced:

$$\dot{x} = (a-1)x - \gamma x^2 . \tag{1.47}$$

The equilibria of (1.47) correspond to the fixed points of (1.16), namely $x_* = 0$ and $x_* = (a-1)/\gamma$. Instead, the study of positive invariance is much more simple: indeed, the set of biological significance $\{x \in \mathbf{R} : x \geq 0\}$ is the union of four orbits (the two equilibrium points, the interval $(0, (a-1)/\gamma)$ and the half-line $\{x > (a-1)/\gamma\}$) and so it is actually invariant.

Note that contrary to what happens in the discrete case, periodic orbits cannot arise for the model (1.47). ∎

Example 1.35 A *predator-prey* model aims to describe the evolution of two (in the simplest case) populations living in the same environment. The prey population, denoted by x, constitutes the food of the predator population, denoted by y. We assume that when separation is maintained between the two populations, the preys growth according to Malthus' law, while the predators decrease until extinction. Any encounter between a prey and a predator gives rise, with some probability, to an advantage for the predator population, and a disadvantage for the prey population. Assume that a unit of time has been fixed, and that the number of individuals of each population is surveyed at integers multiples of such unit. If the environment where such populations live is bounded and if they are homogeneously distributed, it seems reasonable to assume that the frequency of the encounters is measured by the product xy (*homogeneous mixing* assumption). We are so led to write the following mathematical model in $\mathbf{R}^2$, which will be referred to as the discrete time *Lotka-Volterra* model:

$$\begin{cases} x_{n+1} = ax_n - px_n y_n \\ y_{n+1} = by_n + px_n y_n . \end{cases} \tag{1.48}$$

Here, a, b denote respectively the incremental factors of preys and predators ($a > 1, 0 < b < 1$, compare with Example 1.2), while $p \geq 0$ denotes the so-called *factor of interspecific competition*. System (1.48) admits two fixed points: the origin of $\mathbf{R}^2$ and, if $p > 0$, the point $((1-b)/p, (a-1)/p)$.

From a biological point of view, system (1.48) makes sense when $x \geq 0, y \geq 0$. Unfortunately, the quadrant $x \geq 0, y \geq 0$ is not positively invariant for (1.48): indeed, if for some n, $x_n > 0$ and $y_n > a/p$, next iteration will leads $x_{n+1} < 0$.

Under the proportional increment hypothesis, the discrete model (1.48) yields the classical Lotka-Volterra continuous time model, that is

$$\begin{cases} \dot{x} = rx - pxy \\ \dot{y} = -sy + pxy \end{cases} \tag{1.49}$$

where $r = a - 1 > 0$ and $s = 1 - b \in (0, 1)$. Note that for (1.49), the quadrant $x \geq 0$, $y \geq 0$ is actually positively invariant. The equilibrium points correspond, even in this case, to the fixed points of (1.48): the origin and $(s/p, r/p)$. ∎

Example 1.36 One of the most popular mathematical model in epidemiology is the so called *SIR* model (Kermack and Mc Kendrick model). It belongs to the class of compartmental models, which analyze the behavior of a population partitioned into several subsets, called compartments. While the number of individuals of the population remains constant, individuals can move from a compartment to another.

Although the predictions of the SIR model are strongly dependent on the statistical identification of the parameters (very difficult to establish precisely in advance), and obtained under restrictive assumptions and simplifications, it allows us to give an impressive representation, even if merely qualitative, about possible evolutions of an epidemic, and to simulate what happens for small variations of the parameters.

Let N be the number of individuals of the population. The partition consists of three compartments: the set of susceptibles (individuals healthy but exposed to infection), the set of infected (individuals ill and infectious) and the set of restored (including deceased). Assume that we start to monitor the evolution of the epidemic starting at the time $t = 0$. Let us denote respectively by x_n the number of susceptibles and by y_n the number of infected, after n units of time. The number of restored can be obtained by difference (being the overall population constant) and therefore there is no need to take into account of them in the model. Let $s \in (0, 1)$ be the rate of individuals restored from illness during each unit of time. In the discrete time version, the model writes

$$\begin{cases} x_{n+1} = x_n - px_n y_n \\ y_{n+1} = y_n - sy_n + px_n y_n \end{cases} \tag{1.50}$$

with $0 < b = 1 - s < 1$, $p \geq 0$. This system can be thought of as a degenerate form of the Lotka-Volterra system. It has infinitely many fixed points: more precisely, every point of the form $(x, 0)$, and only these points, are fixed points.

Although system (1.50) is mathematically defined on the whole of $\mathbf{R}^2$, from the point of view of the application we are interested in we can clearly limit ourselves to the triangle $T = \{(x, y) : x \geq 0, y \geq 0, x + y \leq N\}$.

Claim. Assume that

$$p < \frac{1}{N}. \tag{1.51}$$

Then, the triangle T is positively invariant, that is

$$(x, y) \in T \implies (x^+, y^+) \in T.$$

Proof of the claim From $x \geq 0$, $y \geq 0$ it readily follows that $y^+ \geq 0$ and $x^+ \leq N$. The statement $x^+ \geq 0$ is equivalent to $y \leq 1/p$, and this is true in consequence of (1.51), since $y \leq N$. It remains to prove that $y^+ \leq N - x^+$, namely

$$by + pxy \leq N - x + pxy$$

which reduces to

$$y \leq \frac{N}{b} - \frac{x}{b} .$$

This last inequality follows from $y \leq N - x$, since $b < 1$. ∎

Other information about the qualitative behavior of the solutions of (1.50) can be easily obtained. If the initial state $(\hat{x}, \hat{y}) \in T$, the first equation shows that the sequence x_n is decreasing. The sequence y_n is increasing as far as $b + px_n > 1$, then it becomes decreasing. It follows that the number of infected reaches a maximum when x_n is around the value $(1 - b)/p$.

For the simulations of Fig. 1.11 the following re-scaled model has been considered

$$\begin{cases} \tilde{x}_{n+1} = \tilde{x}_n - \tilde{p}\tilde{x}_n\tilde{y}_n \\ \tilde{y}_{n+1} = b\tilde{y}_n + \tilde{p}\tilde{x}_n\tilde{y}_n \end{cases}$$

where $\tilde{x} = x/N$, $\tilde{y} = y/N$ and $\tilde{p} = pN$. Assuming the proportional increment hypothesis, we deduce from (1.50) the well-known continuous time *SIR* model

$$\begin{cases} \dot{x} = -pxy \\ \dot{y} = -sy + pxy . \end{cases} \tag{1.52}$$

The equilibria of (1.52) are on the x-axis and the first quadrant is positively invariant. ∎

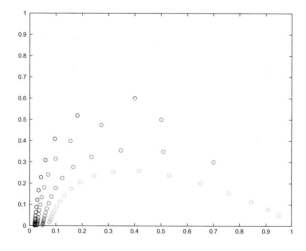

Fig. 1.11 Four trajectories of the re-scaled discrete SIR model (Example 1.36) with $b = 0.7$, $\tilde{p} = 0.91$; each trajectory runs from right to left

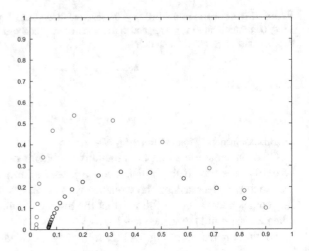

Fig. 1.12 Two trajectories of the re-scaled delayed discrete SIR model (Example 1.37) with $b = 0.7$, $\bar{p} = 0.91$; each trajectory runs from right to left

Example 1.37 The model (1.50) assumes that the incubation and illness average periods both coincide with one unit of time. To take into account of these phenomena in more realistic way, we may introduce a delayed version of (1.50)

$$\begin{cases} x_{n+1} = x_n - px_{n-h}y_{n-h} \\ y_{n+1} = y_n - sy_{n-k} + px_{n-h}y_{n-h} \end{cases} \tag{1.53}$$

where the integers h, k respectively measure, in units of times, the incubation period and the illness period. For instance, with $h = k = 1$, (1.53) can be rewritten as a four order system

$$\begin{cases} u_{n+1} = x_n \\ x_{n+1} = x_n - pu_n v_n \\ v_{n+1} = y_n \\ y_{n+1} = y_n - sv_n + pu_n v_n \end{cases}$$

introducing auxiliary variables $u_n = x_{n-1}$ and $v_n = y_{n-1}$ (see Fig. 1.12).   ∎

References

1. Luenberger, D.G.: Introduction to Dynamic Systems: Theory, Models and Applications. Wiley and Sons, New York (1979)
2. Bernstein, D.S.: Matrix Mathematics: Theory Facts and Formulas with Application to Linear Systems Theory. Princeton University Press, Princeton (2005)
3. LaSalle, J.: The Stability and Control of Discrete Process. Springer, New York (1986)

Chapter 2
Stability and Attraction

The first step in the study of the global behavior of the solutions of a discrete dynamical system

$$x_{n+1} = f(x_n) \tag{2.1}$$

with $f : \Omega \to \mathbf{R}^d$ ($\Omega \subseteq \mathbf{R}^d$ and nonempty), consists in the analysis of the local stability[1] properties of its fixed and periodic points. Stability is a composite notion: we start by some formal definitions enlightening its various aspects. Then we introduce the Lyapunov method. Some room is devoted to the problem of instability. The study of instability is often limited, in many handbooks, to some versions of Chetaev Theorem. Actually, instability may occur in a variety of situations and forms, that we aim to classify and characterize.

After focusing on the case of linear systems, we finally discuss the linearization approach. The reader will recognize analogies and differences with the corresponding definitions relative to equilibrium positions and periodic orbits of differential systems.

The material of this chapter is exposed under the following standing assumptions:

1. Ω is an open subset of $\mathbf{R}^d$;
2. the function f in (2.1) is continuous.

A slightly more general framework could be considered assuming, as in Chap. 1, that Ω is any nonempty subset of $\mathbf{R}^d$ endowed with the induced topology. This would imply some complications in notation and in proofs. Of course, we agree also that $f(\Omega) \subseteq \Omega$, so that the existence of solutions is guaranteed (compare with (1.5)).

---

[1] Throughout this book the term "stability" is always intended in Lyapunov's sense.

© The Author(s), under exclusive license to Springer Nature Switzerland AG 2022
A. Bacciotti, *Discrete Dynamics*, Mathematical Engineering,
https://doi.org/10.1007/978-3-030-95092-7_2

2.1 Main Definitions

2.1.1 Stability of Fixed Points

The following definition emphasizes the fact that stability is a topological property.

Definition 2.1 Let x_* be a fixed point of system (2.1). We say that x_* is *stable* if for each neighborhood $U \subseteq \Omega$ of x_* there exists a neighborhood $V \subset U$ of x_* such that for any initial state $\hat{x}$,

$$\hat{x} \in V \implies f^{[n]}(\hat{x}) \in U$$

for all $n \in \mathbf{N}$. If x_* is not stable, we say that it is *unstable*.

In the previous definition, it is not restrictive to assume that U and V are bounded.

Remark 2.1 Sometimes it may be more convenient to reformulate Definition 2.1 in terms of spherical neighborhoods. Namely, x_* is *stable* if for each $\varepsilon > 0$ there exists $\delta > 0$ such that for any initial state $\hat{x}$,

$$\hat{x} \in \Omega \cap \mathcal{B}(\delta, x_*) \implies f^{[n]}(\hat{x}) \in \mathcal{B}(\varepsilon, x_*) \quad \text{for all} \ \ n \in \mathbf{N} \ . \tag{2.2}$$

∎

Definition 2.2 The *attraction set* of the fixed point x_* is $\mathcal{A}(x_*) = \{\hat{x} \in \Omega : \lim_{n \to +\infty} f^{[n]}(\hat{x}) = x_*\}$.

Note that $\mathcal{A}(x_*) \neq \emptyset$ (it contains at least x_*).

Definition 2.3 We say that the fixed point x_* of system (2.1) is *locally attractive* if x_* is an interior point of its own attraction set i.e., $x_* \in \text{Int } \mathcal{A}(x_*)$. We say that x_* is *globally attractive* if $\mathcal{A}(x_*) = \Omega$.

Note that if x_* is locally attractive, the interior of $\mathcal{A}(x_*)$ is nonempty. In fact, the following proposition holds.

Proposition 2.1 *If the fixed point x_* is locally attractive, then $\mathcal{A}(x_*)$ is open.*

Proof According to the assumption, there exists a neighborhood U of x_* contained in $\mathcal{A}(x_*)$. Now, let us fix an arbitrary $\hat{x} \in \mathcal{A}(x_*)$ and let $\{x_n\}$ be the solution for which $x_0 = \hat{x}$ i.e., $x_n = f^{[n]}(\hat{x})$. By definition, there exists an integer $m \geq 0$ such that for all $n > m$, $x_n \in U$. Fix one $\bar{n} > m$, and let V be a neighborhood of $x_{\bar{n}}$ such that $V \subset U$. Since the map $f^{[\bar{n}]}$ is continuous, the pre-image of V by $f^{[\bar{n}]}$ is a neighborhood W of $\hat{x}$. For each initial state $\hat{y} \in W$, $\hat{z} = f^{[\bar{n}]}(\hat{y}) \in V \subset U$. Hence, $\lim_{\nu \to +\infty} f^{[\nu]}(\hat{z}) = x_*$, and so also

$$\lim_{n \to +\infty} f^{[n]}(\hat{y}) = \lim_{\nu \to +\infty} f^{[\nu+\bar{n}]}(\hat{y}) = \lim_{\nu \to +\infty} f^{[\nu]}(f^{[\bar{n}]}(\hat{y})) = \lim_{\nu \to +\infty} f^{[\nu]}(\hat{z}) = x_* .$$

∎

In general (and contrary to what happens in the continuous time case), $\mathcal{A}(x_*)$ is not necessarily connected (see later, Example 2.15). Further properties and facts about the attraction set can be found in [1], p. 84. The following examples aim to illustrate Definitions 2.1 and 2.3 and to emphasize their independence.

Example 2.1 For the scalar linear system $x^+ = ax$ with $|a| = 1$, $\Omega = \mathbf{R}$ (compare with Examples 1.2 and 1.7) the origin is a stable but not attractive fixed point. For $a = 1$ there are actually infinitely many fixed points which are stable but not attractive. For $|a| < 1$ the origin is stable and (globally) attractive. ∎

Example 2.2 The two-dimensional system of Examples 1.3 and 1.24 has a unique fixed point (the origin). It is clear that this fixed point is unstable. However, its attraction set is a straight line (namely, the line corresponding to the solutions with $c_1 = 0$ in (1.30)) and thus, it contains points distinct from the origin, but its interior is empty. Hence, the origin is not attractive. ∎

Example 2.3 The construction of a system with a fixed point which is attractive but not stable is more difficult and involves homoclinic orbits. We begin by recalling Example 1.19, where we presented a system on the circle C such that for each initial angle $\hat{\theta} \neq 0$, the corresponding solution $\{\theta_n\}$ satisfies $\lim_{n \to -\infty} \theta_n = 0 = \lim_{n \to +\infty} \theta_n$.

The equation (1.23) of Example 1.19 can be coupled with the equation

$$\rho^+ = \sqrt{\rho} \quad (\rho \geq 0) \tag{2.3}$$

to define a system in $\mathbf{R}^2$ written in polar coordinates. Clearly, the solutions of (2.3) write

$$\rho_n = \sqrt[2^n]{\hat{\rho}} .$$

The resulting system has two fixed points: the origin and the point $\rho_* = 1$, $\theta_* = 0$. The latter is attractive (see Fig. 2.1) and actually, the set of attraction is the whole of $\mathbf{R}^2$ except the origin. However, it is clear that it is not stable. ∎

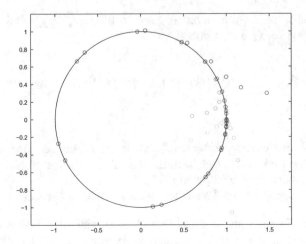

Fig. 2.1 Trajectories for Example 2.3

Example 2.4 It is straightforward to check that the trajectories of the one dimensional system

$$x^+ = f(x) = \begin{cases} \dfrac{x}{2} & \text{if } x \le 0 \\[2mm] 2x & \text{if } x > 0 \end{cases}$$

converge to the origin if the initial state is negative, and diverge to $+\infty$ if the initial state is positive. In a similar situation, a fixed point is sometimes qualified as *semi-stable*. ∎

Example 2.5 The rescaled logistic equation with $a = 1$ and state space extended to $\mathbf{R}$

$$x^+ = x(1 - x)$$

has a unique fixed point for $x = 0$. This point is semi-stable: trajectories starting from a sufficiently small neighborhood of the origin diverge if the initial state is negative and converge in the opposite case (see Fig. 2.2). ∎

Definition 2.4 We say that the fixed point x_* of system (2.1) is [locally/globally] *asymptotically stable* if it is stable and [locally/globally] attractive.

Remark 2.2 If the fixed point x_* is locally asymptotically stable, then all the points in the set of attraction $\mathcal{A}(x_*)$ are uniformly attracted by x_*. This means that for each compact set $K \subset \mathcal{A}(x_*)$ and for each neighborhood U of x_*, there exists an integer $m > 0$ such that for each $\hat{x} \in K$ and each $n > m$, one has

Fig. 2.2 Graphs of solutions of Example 2.5

$$x_n \in U$$

where $\{x_n\}$ is the solution such that $x_0 = \hat{x}$. We emphasize that m depends on K and U but not on $\hat{x}$ (as far as it belongs to K). For a proof, we refer to [1], p. 86.   ■

The following propositions report some interesting facts about stability and attraction.

Proposition 2.2 *The stability property and the attraction property are invariant under topological equivalence.*

Proof We limit ourselves to the stability property. Recall the notation introduced in Sect. 1.1.10, and recall also that under a topological equivalence, a fixed point x_* of (2.1) is mapped into a fixed point y_* of the transformed system $y^+ = g(y)$ (Remark 1.3). Let $U' \subset \Omega'$ be a neighborhood of y_*, and let $U = h(U')$. If x_* is stable, there is a neighborhood $V \subset U$ such that $\hat{x} \in V$ implies $x_n \in U$ for each n, where x_n is the solution of the system (2.1) such that $x_0 = \hat{x}$. Now let $V' = h^{-1}(V)$, and pick any point $\hat{y} \in V'$. We have $\hat{x} = h(\hat{y}) \in V$ and hence for the corresponding solution we have $x_n \in U$. This implies in turn that $h^{-1}(x_n) \in U'$, for each n. To complete the proof, it is sufficient to remark that $y_n = h^{-1}(x_n)$ coincides with the solution of the transformed system issuing from $\hat{y}$.   ■

Proposition 2.3 *Let x_* be a fixed point of system (2.1). Then, x_* is a fixed point of the system*

$$\xi^+ = f^{[2]}(\xi) \tag{2.4}$$

as well. Moreover, x_* is stable for system (2.1) if and only if it is stable for (2.4). Finally, x_* is attractive for system (2.1) if and only if it is attractive for (2.4).

Proof The first statement is obvious. It is also clear that if x_* is stable [attractive] for system (2.1) then it is stable [attractive] also for (2.4), since the solutions of (2.4) are subsequences of the solutions of (2.1). Thus, it remains to prove the converse, that is: if x_* is stable [attractive] for system (2.4) then it is stable [attractive] also for (2.1). Let us consider first the attractivity property. By assumption, there exists a neighborhood U of x_* such that

$$\hat{\xi} \in U \Longrightarrow \lim_{n \to \infty} \xi_n = x_*$$

where ξ_n is the solution of (2.4) with the assigned initial state $\xi_0 = \hat{\xi}$. Let $\varepsilon > 0$ be such that $\mathcal{B}(\varepsilon, x_*) \subset U$. Being f continuous, there exists δ such that $0 < \delta < \varepsilon$ and

$$|x - x_*| < \delta \Longrightarrow |f(x) - f(x_*)| = |f(x) - x_*| < \varepsilon. \tag{2.5}$$

Now let $\hat{x} \in \mathcal{B}(\delta, x_*)$ and let $\{x_n\}$ and $\{y_n\}$ be the solutions of, respectively, (2.1) and (2.4) such that $x_0 = y_0 = \hat{x}$. Clearly, $y_n = x_{2n}$ for each $n = 0, 1, 2, \ldots$. Since $\hat{x} \in U$, by assumption

$$\lim_{n \to \infty} y_n = \lim_{n \to \infty} x_{2n} = x_*. \tag{2.6}$$

Let finally $\{z_n\}$ be the solution of (2.4) such that $z_0 = x_1$. According to (2.5), $x_1 \in U$, which yields

$$\lim_{n \to \infty} z_n = x_*.$$

But for each $n = 0, 1, 2, \ldots$, $z_n = x_{2n+1}$, and hence

$$\lim_{n \to \infty} x_{2n+1} = x_*. \tag{2.7}$$

In conclusion, (2.6) and (2.7) together imply that for each $\bar{x} \in \mathcal{B}(\delta, x_*)$, $\lim_{n \to \infty} x_n = x_*$. As far as the stability property is concerned, the proof is similar. ∎

The previous proposition can be extended to the subsequent iterations of the map f, with some complications in the proof.

Remark 2.3 Given a system of ordinary differential equations in $\mathbf{R}^d$

$$\dot{x} = g(x) \tag{2.8}$$

with $g \in C^1$, we already considered (see Sect. 1.4.2) the associated sampled version

$$x^+ = f(x) \tag{2.9}$$

where $f(x) = \phi(\tau, x)$, and noticed that if x_* is an equilibrium point of (2.8) then x_* is also a fixed point of (2.9). Moreover, x_* is (locally or globally) asymptotically stable as an equilibrium point of (2.8) if and only if it is (locally or globally) asymptotically stable as a fixed point of (2.9). This is proven, for instance, in [2] pp. 123–124. ∎

2.1.2 Stability of Periodic Points

The definitions given in the previous sections can be extended to cycles i.e., orbits of nontrivial periodic solutions, in the following way. If $x \in \mathbf{R}^d$ and M is a nonempty subset of $\mathbf{R}^d$, we say that U is a neighborhood of M if $M \subset \operatorname{Int} U$. Recall that $d(x, M) = \inf_{y \in M} \{|x - y|\}$.

Definition 2.5 Let Γ be a cycle of system (2.1). We say that Γ is *stable* (in Lyapunov's sense) if for each neighborhood U of Γ there exists a neighborhood $V \subset U$ of Γ such that $\hat{x} \in V$ implies $f^{[n]}(\hat{x}) \in U$ for each $n \in \mathbf{N}$. Equivalently, if for each $\varepsilon > 0$ there exists $\delta > 0$ such that for initial state $\hat{x}$,

$$d(\hat{x}, \Gamma) < \delta \implies d(f^{[n]}(\hat{x}), \Gamma) < \varepsilon$$

for all $n \in \mathbf{N}$.

Definition 2.6 The set $\mathcal{A}(\Gamma) = \{\hat{x} \in \mathbf{R}^d : \lim_{n \to +\infty} d(f^{[n]}(\hat{x}), \Gamma) = 0\}$ is called the *attraction set* of the cycle Γ.

Definition 2.7 We say that the cycle Γ of system (2.1) is *attractive* if $\Gamma \subset \operatorname{Int} \mathcal{A}(\Gamma)$.

For the scalar linear system $x^+ = ax$ with $a = -1$ there are actually infinitely many periodic points of period 2. The corresponding cycles are stable but not attractive.

Definition 2.8 We say that the cycle Γ of system (2.1) is *asymptotically stable* if it is stable and attractive.

The definitions above can be given, more generally, replacing Γ by any positively invariant compact set M, and can be also extended to the case where the state space is merely a topological space. Proposition 2.1 can be extended to attractive compact sets, and so also to cycles ([1], p. 84).

Proposition 2.4 *If the cycle Γ of system (2.1) is attractive, then the set of attraction $\mathcal{A}(\Gamma)$ is open.*

In the remaining part of this subsection, we will show how the investigation of the stability properties of a cycle Γ can be reduced to the investigation of the corresponding properties of a point whatsoever belonging to Γ. To simplify the exposition, we limit ourselves to nontrivial cycles of period 2. The extension to an arbitrary period k can be carried on with similar reasoning, but with notational complications. Therefore, in what follows, we consider a pair of points $p, q \in \Omega$ and we agree that:

$$q = f(p), \quad p = f(q), \quad \Gamma = \{p, q\}, \quad p \neq q. \tag{2.10}$$

The points p, q are periodic of period 2 for (2.1), and hence fixed points for the system

$$x^+ = f^{[2]}(x). \tag{2.11}$$

Next lemma states that under these conditions, p and q share the same stability properties.

Lemma 2.1 *Let the system (2.1) be given, and assume that (2.10) holds.*

(i) *If p is stable as a fixed point of (2.11) then also q is stable as a fixed point of (2.11).*

(ii) *if p is attractive as a fixed point of (2.11) then also q is attractive as a fixed point of (2.11).*

Proof First of all, we remark that if $\{x_n\}$ is a solution of (2.11) then $\{y_n\}$, where $y_n = f(x_n)$, is a solution of (2.11) as well. Indeed,

$$y_{n+1} = f(x_{n+1}) = f(f^{[2]}(x_n)) = f^{[2]}(f(x_n)) = f^{[2]}(y_n) \ .$$

Let us consider statement (i). Let $\varepsilon > 0$ be fixed, and let $\varepsilon' > 0$ be such that

$$|y - p| < \varepsilon' \Longrightarrow |f(y) - q| < \varepsilon \ .$$

Such an ε' exists since f is continuous. Then, by the stability assumption, we can find a positive number δ' such that

$$|\hat{y} - p| < \delta' \Longrightarrow |y_n - p| < \varepsilon'$$

for each n, where $\{y_n\}$ is the solution of (2.11) for which $y_0 = \hat{y}$. Finally, using again the continuity of f, we find $\delta > 0$ in such a way that

$$|x - q| < \delta \Longrightarrow |f(x) - p| < \delta' \ .$$

It is not restrictive to assume $\delta' < \varepsilon'$ and $\delta < \varepsilon$. Now pick any $\hat{x} \in \mathcal{B}(\delta, q)$, and consider the solution $\{x_n\}$ of (2.11) such that $x_0 = \hat{x}$. The concatenation of the previous inequalities shows that $|x_n - q| < \varepsilon$ for each n, as required.

Now let us prove statement (ii). By assumption, there exists $\varepsilon > 0$ such that for each $\hat{y} \in \mathcal{B}(\varepsilon, p)$, $\lim_{n \to +\infty} y_n = p$, where $\{y_n\}$ is the solution of (2.11) such that $y_0 = \hat{y}$. Let $\delta > 0$ be such that

$$|x - q| < \delta \Longrightarrow |f(x) - p| < \varepsilon \ .$$

Such a δ exists since f is continuous. Hence, for each $\hat{x} \in \mathcal{B}(\delta, q)$, we have $|f(\hat{x}) - p| < \varepsilon$ and so $\lim_{m \to +\infty} f^{[2m]}(f(\hat{x})) = p$. Using again the continuity of f, it follows

$$\lim_{m \to +\infty} f(f^{[2m]}(f(\hat{x}))) = f(p)$$

which is the same as

$$\lim_{m \to +\infty} f^{[2m+2]}(\hat{x}) = \lim_{m \to +\infty} f^{[2m]}(\hat{x}) = q \ .$$

The statement is proven, $\hat{x}$ being an arbitrary point of an open set which contains q. ∎

Proposition 2.5 *Let the system (2.1) be given, and assume that (2.10) holds.*

(i) p is stable as a fixed point of (2.11) if and only if the cycle Γ is stable as an orbit of system (2.1).

(ii) p is attractive as a fixed point of (2.11) if and only if the cycle Γ is attractive as an orbit of system (2.1).

Proof Let us prove statement (i), assuming first that p is a stable fixed point of system (2.11). By Lemma 2.1, the same holds for q. Then, for each fixed $\varepsilon > 0$ we can find a positive number δ_p such that

$$|\hat{y} - p| < \delta_p \Longrightarrow |y_n - p| < \varepsilon \tag{2.12}$$

and a positive number δ_q such that

$$|\hat{y} - q| < \delta_q \Longrightarrow |y_n - q| < \varepsilon \tag{2.13}$$

for each n, where $\{y_n\}$ represents the solution of (2.11) for which $y_0 = \hat{y}$. Since f is continuous, there exists $\delta'_p > 0$ such that

$$|x - p| < \delta'_p \Longrightarrow |f(x) - f(p)| = |f(x) - q| < \delta_q \ . \tag{2.14}$$

Without loss of generality, we can assume that $\delta'_p < \delta_p$. Now, let $\hat{x}$ be such that $|\hat{x} - p| < \delta'_p$ and let $\{x_n\}$ be the solution of (2.1) for which $x_0 = \hat{x}$. Clearly, $\{x_{2m}\}$ is the solution of (2.11) with the same initial state, so according to (2.12), we have

$$|x_{2m} - p| < \varepsilon \tag{2.15}$$

for each m. On the other hand, (2.14) yields $|f(\hat{x}) - f(p)| = |x_1 - q| < \delta_q$. Moreover, $\{x_{2m+1}\}$ is the solution of (2.11) corresponding to the initial state x_1 and hence, by (2.13), we have

$$|x_{2m+1} - q| < \varepsilon \tag{2.16}$$

for each m. Summing up, if $|\hat{x} - p| < \delta'_p$ then either $|x_n - p| < \varepsilon$ (when n is even) or $|x_n - q| < \varepsilon$ (when n is odd). The argument can be repeated for the point q, which provides us a number δ'_q. The number δ to be used to check the stability of $\Gamma = \{p, q\}$ is computed as the minimum between δ'_p and δ'_q.

Vice-versa, assume that Γ is stable for system (2.1). We will prove that p is stable for system (2.11) (the same arguments can be repeated for q). Let $\varepsilon > 0$ be fixed in such a way that $\mathcal{B}(\varepsilon, p) \cap \mathcal{B}(\varepsilon, q) = \emptyset$. By the continuity of f, there exists $\tilde{\varepsilon} > 0$ such that

$$|y - p| < \tilde{\varepsilon} \implies |f^{[2]}(y) - f^{[2]}(p)| = |f^{[2]}(y) - p| < \varepsilon . \tag{2.17}$$

We may assume that $\tilde{\varepsilon} < \varepsilon$. Then by the stability assumption, there exists $\delta < \tilde{\varepsilon}$ such that from $|\hat{x} - p| < \delta$ it follows, for each n, either $|x_n - p| < \tilde{\varepsilon}$ or $|x_n - q| < \tilde{\varepsilon}$. In order to prove the statement, we need to prove that from $|\hat{x} - p| < \delta$ it actually follows $|x_{2n} - p| < \varepsilon$ for each n. If $n = 0$ this is true by construction. We can complete the proof by contradiction. Assume that for some n we have $|x_{2n} - p| < \tilde{\varepsilon}$ but $|x_{2n+2} - p| \geq \tilde{\varepsilon}$. Then it must be $|x_{2n+2} - q| < \tilde{\varepsilon}$ and so

$$|x_{2n+2} - p| \geq \varepsilon . \tag{2.18}$$

On the other hand, by (2.17)

$$|x_{2n} - p| < \tilde{\varepsilon} \implies |f^{[2]}(x_{2n}) - f^{[2]}(p)| < \varepsilon$$

which it turns gives

$$|f^{[2]}(x_{2n}) - p| = |x_{2n+2} - p| < \varepsilon$$

and we have a contradiction to (2.18).

Now we consider statement (ii), assuming first that p is attractive as a fixed point of (2.11). Let $\hat{x}$ be a point belonging to the set of attraction of p (thought of as a fixed point of system (2.11)), and let $\{x_n\}$ be the solution of system (2.1) such that $x_0 = \hat{x}$. Then the sequence $\{x_{2n}\}$ is a solution of system (2.11) and it converges to p, namely we have

$$\lim_{k \to +\infty} x_{2k} = p . \tag{2.19}$$

On the other hand, f is continuous and so

$$\lim_{n \to +\infty} x_{2n+1} = \lim_{n \to +\infty} f(x_{2n}) = f(p) = q. \tag{2.20}$$

Putting together (2.19) and (2.20) we see that $\lim_{n \to +\infty} d(x_n, \Gamma) = 0$. This shows that there exists an open neighborhood of p which is contained in the set $\mathcal{A}(\Gamma)$. By virtue of Lemma 2.1, the same argument can be repeated for q.

Vice-versa, assume that $\Gamma = \{p, q\}$ is attractive for system (2.1) and let $\hat{x}$ a point in the set of attraction of Γ, $\mathcal{A}(\Gamma)$. Let us fix $\varepsilon > 0$ in such a way that $\mathcal{B}(\varepsilon, p) \cap \mathcal{B}(\varepsilon, q) = \emptyset$ and $\mathcal{B}(\varepsilon, p) \cup \mathcal{B}(\varepsilon, q) \subset \mathcal{A}(\Gamma)$. Then, by continuity, we can find positive numbers $\varepsilon_p, \varepsilon_q$ such that

$$|y - p| < \varepsilon_p \implies |f^{[2]}(y) - p| < \varepsilon \quad \text{and} \quad |y - q| < \varepsilon_q \implies |f^{[2]}(y) - q| < \varepsilon .$$
$$(2.21)$$

Finally, let $\tilde{\varepsilon} = \min\{\varepsilon_p, \varepsilon_q\}$. Without loss of generality, we can assume that $\tilde{\varepsilon} < \varepsilon$. Now fix a positive $\sigma < \tilde{\varepsilon}$. By the attractivity assumption, there is an integer $\tilde{n}$ such that from $n \geq \tilde{n}$ implies

$$x_n \in \mathcal{B}(\sigma, p) \cup \mathcal{B}(\sigma, q) \qquad (2.22)$$

where $\{x_n\}$ is the solution of (2.1) for which $x_0 = \hat{x}$ (note that $\mathcal{B}(\sigma, p)$ and $\mathcal{B}(\sigma, q)$ are disjoint by construction). We may assume, without restriction, that $\tilde{n}$ is even i.e., $\tilde{n} = 2\tilde{m}$. For the given point $\hat{x}$, we so have two cases to be distinguished. Assume it happens that

$$x_{\tilde{n}} \in \mathcal{B}(\sigma, p) .$$

We will prove that, in this case, for each $m \geq \tilde{m}$ one has $x_{2m} \in \mathcal{B}(\sigma, p)$. We argue by contradiction, assuming that for some $m_1 \geq \tilde{m}$,

$$|x_{2m_1} - p| < \sigma \quad \text{but} \quad |x_{2m_1+2} - p| \geq \sigma.$$

Then necessarily, by (2.22) it must be $|x_{2m_1+2} - q| < \sigma$, which in turn implies $|x_{2m_1+2} - p| > \varepsilon$. This is a contradiction to (2.21). Since σ is arbitrary, we have so proven that $\lim_{m \to +\infty} f^{[2m]}(\hat{x}) = p$.

In the other case where $x_{\tilde{n}} \in \mathcal{B}(\sigma, q)$, we apply the analogous reasoning and will arrive to the conclusion that $\lim_{m \to +\infty} f^{[2m]}(\hat{x}) = q$.

Thus we see that $\mathcal{A}(\Gamma)$ can be decomposed as the disjoint union of two sets $\mathcal{A}_1$, $\mathcal{A}_2$ which constitute, respectively, the sets of attraction of p and q (say, $p \in \mathcal{A}_1$, $q \in \mathcal{A}_2$). Since $\mathcal{A}(\Gamma)$ is open, both $\mathcal{A}_1$, $\mathcal{A}_2$ are open. ∎

Remark 2.4 As a consequence of the previous proof, we see also that $\mathcal{A}(\Gamma)$ is not connected for any cycle of period 2. For instance, in the case of the logistic equation, the connected components of $\mathcal{A}(\Gamma)$ are determined later (see Example 2.15). ∎

2.2 Uniform Boundedness

The notion of uniform boundedness (sometimes referred to as Lagrange stability) concerns the behavior of the solutions for large values and it is not directly related to some special solution like a fixed point or a cycle.

Definition 2.9 We say that system (2.1) has the property of *uniform boundedness* (or that (2.1) is *uniformly bounded*) if for each $S > 0$ there exists $R > S$ such that from $|\hat{x}| < S$ it follows $|x_n| < R$ for each $n \in \mathbf{N}$, where $\{x_n\}$ is the solution for which $x_0 = \hat{x}$.

If (2.1) is uniformly bounded, then clearly every solution is bounded, but the converse is not true in general. Symmetric to uniform boundedness, we may also introduce a further notion.

Definition 2.10 We say that system (2.1) is *bounded away* from the origin if for each $\sigma > 0$ there exists $\rho < \sigma$ such that from $|\hat{x}| > \sigma$ it follows $|x_n| > \rho$ for each $n \in \mathbf{N}$, where $\{x_n\}$ is the solution for which $x_0 = \hat{x}$.

A system may be bounded away from the origin even if the origin is not a fixed point (this happens for instance with the scalar system $x^+ = 1$). On the other hand, the previous property does not prevent the origin from being a stable fixed point, as we can see for instance looking at the scalar system $x^+ = -x$. Definition 2.10 is referred to the origin, but this is not essential. Analogous definitions can be given with respect to any other point or bounded region of $\mathbf{R}^d$.

2.3   Lyapunov Functions

The Lyapunov functions method is one of the most popular method in stability analysis. Inspired by Lagrange's studies about classical mechanics, it was formalized for the case of differential equations by A. M. Lyapunov at the end of nineteenth century. It can be applied also to discrete dynamical systems. Here, we follow essentially the exposition in [3].

Recall that a function $V(x) : \Omega \to \mathbf{R}$ is:

- *positive definite* at a point $p \in \Omega$ if there exists a neighborhood $U \subset \Omega$ of p such that $V(p) = 0$ and $V(x) > 0$ for each $x \in U, x \neq p$;
- *positive semidefinite* at a point $p \in \Omega$ if there exists a neighborhood $U \subset \Omega$ of p such that $V(p) = 0$ and $V(x) \geq 0$ for each $x \in U$;
- *negative definite* at a point $p \in \Omega$ if $-V(x)$ is positive definite at p;
- *negative semidefinite* at a point $p \in \Omega$ if $-V(x)$ is positive semidefinite at p;
- *indefinite* at a point $p \in \Omega$ if $V(x)$ takes both positive and negative values in each neighborhood $U \subset \Omega$ of p.

To any pair of functions $V(x) : \Omega \to \mathbf{R}$ and $f(x) : \Omega \to \mathbf{R}^d$ for which $f(\Omega) \subseteq \Omega$, we associate a new function $V^+(x) : \Omega \to \mathbf{R}$ defined as

$$V^+(x) = V(f(x)) - V(x) .$$

Generally speaking, a *Lyapunov function* for the system (2.1) is a positive definite function $V(x)$ such that $V^+(x)$ is negative definite or semidefinite; the precise requirements about $V(x)$ depends on the problem at hand.

2.3.1 Stability

The "First Lyapunov Theorem" concerns stability.

Theorem 2.1 *Let the system (2.1) with $f(x) : \Omega \to \mathbf{R}^d$ be given, and let x_* be a fixed point. Assume that there exists a continuous function $V(x) : \Omega \to \mathbf{R}$ such that $V(x)$ is positive definite at x_* and $V^+(x)$ is negative semidefinite at x_*. Then, x_* is stable.*

Proof Let $\eta > 0$ be such that $V(x)$ is positive definite and $V^+(x)$ is negative definite at x_* with $U = \mathcal{B}(\eta, x_*)$. Let us fix a number $\varepsilon > 0$, such that $\varepsilon < \eta$; this can be done without loss of generality. Let $m = \min_{x \in L} V(x)$, where $L = \{x : |x - x_*| = \varepsilon\}$. Since V is continuous and L compact, the minimum exists. Moreover, V being positive on L, we have $m > 0$. The set $G = \{x \in \Omega : V(x) < m/2\}$ is open and $x_* \in G$. Let G_0 be the connected component containing x_*. Clearly, $G_0 \subset \mathcal{B}(\varepsilon, x_*) \subset \mathcal{B}(\eta, x_*)$.

If $z \in G_0$, then $V^+(z) \leq 0$, namely

$$V(f(z)) \leq V(z) < \frac{m}{2}$$

which implies that $f(z) \in G$. On the other hand, x_* and z belong to the same connected component of G and since f is continuous, $f(x_*)$ and $f(z)$ must belong to the same connected component of $f(G) \subset G$. But $f(x_*) = x_*$, hence $f(z) \in G_0$.

Taking a positive number δ such that $\mathcal{B}(\delta, x_*) \subset G_0$ and using repeatedly the previous argument, we easily recover the stability property. ∎

Actually, the argument of the previous proof allows us to state the following corollary.

Corollary 1 *Under the assumption of Theorem 2.1, every open neighborhood of the fixed point x_* contains an open positively invariant neighborhood of x_*.*

The converse of Theorem 2.1 is false in general. A counterexample can be given taking a discretization of well known analog counterexamples for the continuous time case (see [1]).

By virtue of Proposition 2.5, the First Lyapunov Theorem can be extended to the case of periodic points. We have for instance the following result.

Corollary 2 *Let $\Gamma = \{p, q\}$ be a cycle of period 2 $(p \neq q)$ for system (2.1) and let $\Omega' \subset \Omega$ be an open set containing Γ such that $f(\Omega') \subseteq \Omega'$. Assume that there exists a continuous function $V(x) : \Omega \to \mathbf{R}$ such that $V(p) = V(q) = 0$, $V(x) > 0$ for $x \in \Omega \setminus \Gamma$. Assume finally that $V^+(x) \leq 0$ for each $x \in \Omega'$. Then, Γ is stable.*

Proof The hypotheses guarantee that V is positive definite and V^+ is negative semidefinite at p. Let $U \subset \Omega'$ be a neighborhood of p such that $V(x) > 0$ $(x \neq p)$ and $V^+(x) \leq 0$ for each $x \in U$. Since $f(x) \in \Omega'$ for each $x \in U$, we have

$$V(f^{[2]}(x)) = V(f(f(x))) \le V(f(x)) \le V(x)$$

for each $x \in U$. Hence, the hypotheses of First Lyapunov Theorem are met for the system

$$x^+ = f^{[2]}(x)$$

at the point p. The conclusion follows by virtue of Proposition 2.5. ∎

2.3.2 Boundedness

As far as the boundedness property introduced in Sect. 2.2 is concerned, we may give the following Lyapunov-like theorem. Recall that an everywhere defined function $V(x) : \mathbf{R}^d \to \mathbf{R}$ is said to be *radially unbounded* if $\lim_{|x| \to +\infty} V(x) = +\infty$.

Theorem 2.2 *Let the system (2.1) with $f(x) : \mathbf{R}^d \to \mathbf{R}^d$ be given. Assume that there exists a continuous, radially unbounded function $V(x) : \mathbf{R}^d \to \mathbf{R}$ such that for each $x \in \mathbf{R}^d$*

$$V(x) > 0 \ (x \ne 0) \ \text{and} \ V^+(x) \le 0 .$$

Then the system is uniformly bounded.

Proof Assume, by contradiction, that there exists $S > 0$ such that for each $R > S$ there exists an initial state $\hat{x}$ and an integer $\bar{n}$ for which

$$|\hat{x}| < S \text{ and } |x_{\bar{n}}| \ge R . \tag{2.23}$$

Let $M = \max_{|x| \le S} V(x) > 0$. By the radial unboundedness of V, there is a number $R_0 > S$ such that $V(x) > 2M$ for each x with $|x| > R_0$. Take $R > R_0$, and let $\hat{x}$ and $\bar{n}$ be as in (2.23). We clearly have $V(\hat{x}) \le M < 2M < V(x_{\bar{n}})$. This is impossible. Indeed, since $V^+(x) \le 0$ everywhere, we have

$$V(\hat{x}) = V(x_0) \ge V(x_1) \ge \ldots \ge V(x_{\bar{n}}) .$$

∎

A lightly more general result is given in [1], p. 14. Next proposition concerns the property of being bounded away from the origin.

Proposition 2.6 *Let the system (2.1) with $f(x) : \Omega \to \mathbf{R}^d$ be given, and let $0 \in \Omega$. Assume that there exists a continuous function $V(x) : \Omega \to \mathbf{R}$ such that for each $x \in \Omega$:*

1. $V(0) = 0$ and $V(x) > 0$ for each $x \in \Omega$, $x \neq 0$;
2. there exist positive numbers η, μ such that $\mathcal{B}(2\eta, 0) \subset \Omega$ and $V(x) \geq \mu$ for each $x \in \Omega \setminus \mathcal{B}(\eta, 0)$;
3. $V^+(x) \geq 0$ for each $x \in \Omega$.

Then the system is bounded away from the origin.

Proof Again, we may argue by contradiction. Assume that there is a number $\sigma > 0$ such that for each $\rho < \sigma$ we can find an initial state $\hat{x}$ and an integer $\bar{n}$ for which

$$|\hat{x}| > \sigma, \ |x_{\bar{n}}| \leq \rho . \tag{2.24}$$

Without restriction, we can take $\sigma < \eta$. Let $c = \min_{|x| \geq \sigma, x \in \Omega} V(x)$. Under our assumption, we have $c > 0$, and by continuity we can find $\rho < \sigma$ such that $V(x) < c/2$ if $|x| < \rho$. Using (2.24), we have for some $\hat{x}$ and $\bar{n}$,

$$V(x_{\bar{n}}) < \frac{c}{2} < c \leq V(\hat{x})$$

while, being $V^+(x) \geq 0$ for each $x \in \Omega$,

$$V(\hat{x}) = V(x_0) \leq V(x_1) \leq \ldots \leq V(x_{\bar{n}}) .$$

We have so found a contradiction.   ∎

We emphasize that to obtain a local result, Proposition 2.6 exploits an assumption of global nature. We will encounter later other similar situations.

2.3.3   Invariance

The so-called "Invariance Principle" is a powerful tool for stability analysis. It rests on the key notion of limit set. The definition of limit set for discrete systems is a natural modification of the analogous definition for continuous time systems (see for instance [4]).

Definition 2.11 Let $\hat{x} \in \Omega$ be given. A point $y \in \Omega$ is said to be a *positive limit point* of $\hat{x}$ with respect to system (2.1) if there exists a sequence of integer numbers $\{n_k\}$ such that $n_k \to +\infty$ and $\lim_{k \to +\infty} x_{n_k} = y$, where $\{x_n\}$ is the solution of (2.1) issued from $\hat{x}$.

The set of all the positive limit points of $\hat{x}$ is called the *positive limit set* of $\hat{x}$, and it is denoted $\Lambda^+(\hat{x})$.

If (2.1) is invertible, one can define also negative limit points and the negative limit set $\Lambda^-(\hat{x})$ in the obvious way. Of course, the positive limit set of a point $\hat{x}$ may be empty. The main properties of the positive limit set are resumed in the following proposition. The reader is referred to [3] p. 6 for the proof.

Proposition 2.7 *If the solution $\{x_n\}$ such that $x_0 = \hat{x} \in \Omega$ is bounded, then $\Lambda^+(\hat{x})$ is nonempty, compact and positively invariant. Moreover,*

$$\lim_{n \to +\infty} d(x_n, \Lambda^+(\hat{x})) = 0 . \tag{2.25}$$

If Γ denotes the orbit of $\{x_n\}$, then $\overline{\Gamma} = \Gamma \cup \Lambda^+(\hat{x})$. Finally, if the system is invertible, $\Lambda^+(\hat{x})$ is also negatively invariant.

Contrary to what happens in the case of continuous time systems, for discrete dynamical systems the limit set need not to be connected, even when it is compact. As an example, we can think to an attractive cycle of period 2. In order to establish the Invariance Principle, our first step is the following Lemma.

Lemma 2.2 *Let G be a nonempty set such that $\overline{G} \subseteq \Omega$ and let $\hat{x} \in G$. Let $V(x) : \Omega \to \mathbf{R}$ be a continuous real function. Let finally $\{x_n\}$ be the solution of system (2.1) such that $x_0 = \hat{x}$ and assume the following facts.*

(i) *There exists an index m_1 such that $x_n \in G$ for all $n \geq m_1$.*
(ii) *The solution $\{x_n\}$ is bounded.*
(iii) *There exists an index m_2 such that the sequence $\{V(x_n)\}_{n \geq m_2}$ is monotone (increasing or decreasing).*

Then, $\Lambda^+(\hat{x}) \subseteq E$, where $E = \{x \in \overline{G} : V^+(x) = 0\}$.

Proof Let Γ be the orbit of $\{x_n\}$. Because of (ii), $\overline{\Gamma}$ is compact. Moreover, (ii) and (i) imply also that $\Lambda^+(\hat{x})$ is nonempty, compact and contained in $\overline{G}$.

Let $y, z \in \Lambda^+(\hat{x})$. By definition, there are sequences of natural numbers $\{v_k\}$ and $\{\mu_k\}$ such that $x_{v_k} \to y$ and $x_{\mu_k} \to z$. Since V is continuous and $x_n \in \overline{\Gamma}$, the sequence $\{V(x_n)\}$ is bounded. Hence by (iii), the limit

$$\lim_{n \to +\infty} V(x_n) = c \in \mathbf{R}$$

exists. Using again the continuity of V, we have

$$\lim_{n \to +\infty} V(x_n) = \lim_{k \to +\infty} V(x_{v_k}) = V(y)$$

and

$$\lim_{n \to +\infty} V(x_n) = \lim_{k \to +\infty} V(x_{\mu_k}) = V(z) .$$

This implies that $V(x)$ is constant and, more precisely, $V(x) = c$ for every $x \in \Lambda^+(\hat{x})$.

To complete the proof, we need to recall that $\Lambda^+(\hat{x})$ is positively invariant. Thus, if $y \in \Lambda^+(\hat{x})$, then also $f(y) \in \Lambda^+(\hat{x})$ and hence $V^+(y) = V(f(y)) - V(y) = c - c = 0$. ∎

Next theorem is a possible formulation of the Invariance Principle.

Theorem 2.3 *Let the system (2.1) with $f : \Omega \to \mathbf{R}^d$ be given, and let G be such that $\overline{G} \subset \Omega$. Let $\hat{x} \in G$, and assume that the orbit of the solution $\{x_n\}$ such that $x_0 = \hat{x}$ is contained in G and bounded.*

Let $V(x) : \Omega \to \mathbf{R}$ be a continuous function, and assume that $V^+(x) = V(f(x)) - V(x) \leq 0$ for each $x \in G$. Let as before $E = \{x \in \overline{G} : V^+(x) = 0\}$. Then, there exists a number c such that

$$\lim_{n \to +\infty} d(x_n, M \cap L_c) = 0 \qquad (2.26)$$

where M is the largest positively invariant set contained in E and $L_c = \{x \in \Omega : V(x) = c\}$.

Proof The assumption of Lemma 2.2 are satisfied. In particular, $\{V(x_n)\}$ turns out to be decreasing by virtue of the assumption $V^+(x) \leq 0$ on G. The number c is identified in the proof of Lemma 2.2. Finally, (2.26) follows from (2.25) and the obvious fact that $\Lambda^+(\hat{x}) \subseteq M$. ∎

With respect to (2.25), the advantage of (2.26) is that in some applications M is easier to determine than $\Lambda^+(\hat{x})$.

2.3.4 Asymptotic Stability

The discrete version of the classical Lyapunov theorem about asymptotic stability can be easily proven on the base of the Invariance Principle.

Theorem 2.4 *Let the system (2.1) with $f : \Omega \to \mathbf{R}^d$ be given, and let x_* be a fixed point. Assume that there exists a continuous function $V(x) : \Omega \to \mathbf{R}$ such that $V(x)$ is positive definite at x_* and $V^+(x)$ is negative definite at x_*. Then, x_* is locally asymptotically stable.*

Proof First of all, we remark that the hypotheses of Theorem 2.1 are met, and hence x_* is stable. Let $\varepsilon > 0$ be such that $\mathcal{B}(\varepsilon, x_*) \subset \Omega$ and, in addition, $V(x) > 0$ and $V^+(x) < 0$ for each $x \neq x_*$, $x \in \mathcal{B}(\varepsilon, x_*)$. Because of stability, we can find $\delta > 0$ such that for each $\hat{x} \in \mathcal{B}(\delta, x_*)$ the corresponding solution $\{x_n\}$ remains in $G = \mathcal{B}(\varepsilon/2, x_*)$ for every n. In particular, it is bounded. Since $V^+(x)$ vanishes only for $x = x_*$ on $\mathcal{B}(\varepsilon, x_*)$, with the notation of Theorem 2.3 we have $E = M = \{x_*\}$. Hence, $\lim_{n \to +\infty} x_n = x_*$. This shows that x_* is attractive. ∎

The previous theorem concerns the case of fixed points. Similar to what we did in Corollary 2 and still based on Proposition 2.5, an analogous result can be stated for the case of periodic points.

The information provided by Theorem 2.4 is merely local. However, under a slightly more restrictive assumption, exploiting the Invariance Principle approach we can obtain estimations of the size of the set of attraction.

Theorem 2.5 *In addition to the assumptions of Theorem 2.4, let us assume that* $V(x) > 0$ *and* $V^+(x) < 0$ *for each* $x \in \Omega$, $x \neq x^*$. *Let* $\ell > 0$, *and consider the sub-level set* $S_\ell = \{x \in \Omega : V(x) < \ell\}$. *If* S_ℓ *is bounded, then* $S_\ell \subset \mathcal{A}(x_*)$.

Proof First we notice that $x_* \in S_\ell$. If $x \in S_\ell$ ($x \neq x_*$), from $V^+(x) < 0$ it follows $V(f(x)) < V(x)$ and hence $V(f(x)) < \ell$. This means that S_ℓ is positively invariant. Therefore, the assumptions of Theorem 2.3 are met with $G = S_\ell$, and hence for each $\hat{x} \in S_\ell$ the corresponding solution converges to x_*.   ∎

The previous result cannot be used in general to ascertain global asymptotic stability. The global version of Theorem 2.4 is provided by the following theorem, where we will make again use of everywhere defined and radially unbounded functions $V(x) : \mathbf{R}^d \to \mathbf{R}$ (see Sect. 2.3.2). The radial unboundedness assumption implies in particular that the sub-level sets of V are bounded and that $\mathbf{R}^d = \bigcup_{\ell>0} S_\ell$.

Theorem 2.6 *In addition to the assumptions of Theorem 2.4, let us assume that* $\Omega = \mathbf{R}^d$ *and that* $V(x)$ *is radially unbounded. Assume further that* $V(x) > 0$ *and* $V^+(x) < 0$ *for each* $x \in \mathbf{R}^d$, $x \neq x^*$. *Then,* x_* *is globally asymptotically stable.*

Contrary to what happens for the First Lyapunov Theorem, Theorems 2.4 and 2.6 are invertible (see for instance [3, 5]).

2.3.5 Instability

Lyapunov functions can be used to ascertain instability, as well. The statement reported below is the discrete version of the classical Chetaev Theorem.

Theorem 2.7 *Let the system (2.1) be given, and let* x_* *be a fixed point. Let* $V(x) :$ $\Omega \to \mathbf{R}$ *be any continuous function such that* $V(x_*) = 0$ *and* $\mathcal{O} = \{x : V(x) > 0\} \neq$ $\emptyset$. *Assume further that* x_* *belongs to the closure of* $\mathcal{O}$, *and* $V^+(x) > 0$ *for each* $x \in \overline{\mathcal{O}} \setminus \{x_*\}$. *Then,* x_* *is unstable.*

Proof Note that under our assumptions $\mathcal{O}$ is open and $x_* \notin \mathcal{O}$. We proceed by contradiction. Let $\varepsilon > 0$ be such that $\overline{\mathcal{B}(\varepsilon, x_*)} \subset \Omega$. If x_* is stable, we can find $\delta > 0$ such that for each $\hat{x} \in \mathcal{B}(\delta, x_*)$ the solution corresponding to $\hat{x}$ satisfies $x_n \in \mathcal{B}(\varepsilon, x_*)$ for each n. In particular, $\{x_n\}$ is bounded. By the assumption about $\mathcal{O}$, we can choose $\hat{x}$ in such a way that $V(\hat{x}) = V(x_0) > 0$. Then we have $V^+(x_0) = V(x_1) - V(x_0) > 0$, that is $V(x_1) > V(x_0) > 0$. This implies in particular $x_1 \in \mathcal{O}$. Using repeatedly the same argument, we find

$$0 < V(x_0) < V(x_1) < V(x_2) \ldots . \tag{2.27}$$

This means that $x_n \in \mathcal{O}$ for each n and that $V(x_n)$ is increasing.

In conclusion, the sequence $\{x_n\}$ mets all the conditions of Lemma 2.2, with $G = \mathcal{B}(\varepsilon, x_*) \cap \mathcal{O}$. On the other hand, since V^+ takes strictly positive values on $\mathcal{O} \setminus \{x_*\}$, we must have $E = \{x_*\}$, so that $\lim_{n\to\infty} x_n = x_*$. This implies in turn, by continuity, that $V(x_n) \to 0$. A contradiction to (2.27). ∎

In order to exclude that the fixed point is attractive we need in general more information about V^+.

Theorem 2.8 *Let x_* be a fixed point of system (2.1). Assume that there exists a continuous function $V(x) : \Omega \to \mathbf{R}$ such that $V(x_*) = 0$ and $V^+(x) > 0$ for each $x \in \Omega$, $x \neq x_*$. Assume also that x_* belongs to the closure of $\mathcal{O} = \{x : V(x) > 0\}$. Then, x_* is not attractive.*

Proof The proof is similar to that of Theorem 2.7. If x_* is attractive, we can find a number $\eta > 0$ such that for each $\hat{x} \in \mathcal{B}(\eta, x_*)$ the corresponding trajectory converges to x_*. We can chose $\hat{x}$ in such a way that $V(\hat{x}) = V(x_0) > 0$. Of course, the sequence $\{x_n\}$ is bounded and there exists a set G such that $x_n \in G$ for each n and $\overline{G} \subset \Omega$. Moreover, by the assumption about V^+, the sequence $\{V(x_n)\}$ is increasing, that is (2.27) holds. We are therefore in a position to apply Lemma 2.2. We conclude $\Lambda^+(\hat{x}) \subseteq E$. But, since x_* is the unique point where $V^+(x) = 0$, we have that $E = \{x_*\}$. Hence, $\lim_{n\to+\infty} x_n = x_*$ and $\lim_{n\to+\infty} V(x_n) = V(x_*) = 0$. This is a contradiction to (2.27), and the statement is proven. ∎

2.4 Repulsivity

Repulsivity is a strong form of instability. In the theory of continuous time differential equations, repulsivity can be intuitively defined in terms of the reversed time system. In other words, we may say that an equilibrium point of the system

$$\dot{x} = g(x) \tag{2.28}$$

is repulsive if it is asymptotically stable for the system

$$\dot{x} = -g(x) . \tag{2.29}$$

Indeed, it is easy to check that if $\phi(t, x)$ is the flow map of system (2.28), then the flow map of system (2.29) is $\phi(-t, x)$. The aim of this section is to propose and to study a possible notion of repulsivity for discrete dynamical systems. But first, we need more insight about the continuous time case.

2.4.1   Continuous Time Case

The characterization of repulsive equilibria is very natural in the case of linear systems: roughly speaking, for a linear differential system defined by a matrix A the origin is repulsive if and only if the eigenvalues of A all have strictly positive real part. For nonlinear systems, the possible occurrence of homoclinic orbits or nested invariant regions (like in the case of the center-focus configuration, see for instance [6] p. 32), complicates things.

In the context of topological dynamics, in [7] p. 65 the notion of *negative asymptotic stability* is introduced. It basically formalizes the aforementioned idea, redefining stability and attraction in the past, instead of in the future. In particular, Theorem 1.1 Chapter VI of [7] provides a complete classification of the local behavior of the flow near an equilibrium point (more generally, of a compact invariant set). Other possible formulations of instability notions are proposed in [4]. We are especially interested in the following result, which suggests a more qualitative interpretation of repulsivity for the continuous time case.

Proposition 2.8 *Consider the differential system in $\Omega \subset \mathbf{R}^d$ defined by (2.28), with $g \in C^1$ and assume that its flow map $\phi(t, x)$ is defined for each $(t, x) \in \mathbf{R} \times \Omega$. Let $g(x_*) = 0$. Then, the following conditions are equivalent.*

(i) *The equilibrium point x_* is locally asymptotically stable for the reversed time system (2.29).*

(ii) *There exists a number $\eta > 0$ such that for each $\hat{x} \neq x_*$, if $|\hat{x} - x_*| < \eta$, then there is a number $T > 0$ such that:*

$$\forall t > T, \quad |\phi(t, \hat{x}) - x_*| \geq \eta .$$

(iii) *There exists a number $\eta > 0$ such that for each $\hat{x} \neq x_*$, if $|\hat{x} - x_*| < \eta$, then there is a number $t > 0$ such that:*

$$|\phi(t, \hat{x}) - x_*| \geq \eta .$$

Proof Without loss of generality, we take $x_* = 0$. Assume that (i) holds, and let $\eta > 0$ be such that $\overline{B(\eta, 0)} \subset \mathcal{A}^-(0) \subset \Omega$, where $\mathcal{A}^-(0)$ denote the set of attraction of the origin for the system (2.29). We argue by contradiction, assuming that (ii) is false. Then for some point $\hat{x} \neq 0$ such that $|\hat{x}| < \eta$, there must exist a sequence $t_n \to +\infty$ such that for each n

$$|x_n| < \eta ,$$

where we set $x_n = \phi(t_n, \hat{x})$.

Since $\overline{B(\eta, 0)}$ is compact, there exist a convergent subsequence $t_{n_k} \to +\infty$ such that

$$x_{n_k} = \phi(t_{n_k}, \hat{x}) \to y \in \Lambda^+(\hat{x}) .$$

We must have $y \neq 0$, otherwise the stability assumption about (2.29) would be contradicted (recall that $\hat{x} \neq 0$). In fact we see that $0 \notin \Lambda^+(\hat{x})$: otherwise, we could find a (different) sequence $s_n \to +\infty$ such that $\phi(s_n, \hat{x}) \to 0$. Again, this is impossible by the stability assumption. On the other hand, we clearly have $|y| \leq \eta$ which implies $y \in \mathcal{A}^-(0)$. Hence, $\lim_{t \to +\infty} \phi(-t, y) = 0$. Since $\Lambda^+(\hat{x})$ is closed and invariant (positively and negatively), we conclude that $0 \in \Lambda^+(\hat{x})$. A contradiction.

The implication (ii) $\Longrightarrow$ (iii) is obvious. Thus, it remains to show that (iii) implies (i). First we consider the stability property. Again, we proceed by contradiction. If the origin is not stable for (2.29), then for some $\varepsilon > 0$ and each $\delta > 0$ we can find $\hat{x} \neq 0$ and $T > 0$ such that $|\hat{x}| < \delta$ and $|\phi(-T, \hat{x})| \geq \varepsilon$. Without loss of generality we can assume[2] $\varepsilon < \eta$. We can therefore construct a sequence of initial states $\hat{x}_i$ and a sequence of positive instants T_i ($i = 1, 2, \ldots$) satisfying the following: $|\hat{x}_i| < \varepsilon$, $\lim_{i \to \infty} \hat{x}_i = 0$ and $|\phi(-T_i, \hat{x}_i)| \geq \varepsilon$. Without loss of generality we can also assume $T_i = \min\{T : |\phi(-T, \hat{x}_i)| \geq \varepsilon\}$, so that for every index i

$$|\phi(-T_i, \hat{x}_i)| = \varepsilon . \tag{2.30}$$

We claim that the sequence $\{T_i\}$ is unbounded. In the opposite case, for $T_0 = \sup T_i > 0$ we should have $T_i \in [0, T_0]$. Since $\hat{x}_i \to 0$, the solutions $\phi(t, \hat{x}_i)$ converge uniformly on $[-T_0, 0]$ to the solution issuing from the origin, which is an equilibrium point. In other words, for arbitrarily small σ and sufficiently large i we should have $|\phi(t, \hat{x}_i)| < \sigma$ for all $t \in [-T_0, 0]$. This is clearly in contrast with (2.30).

Now, let $z_i = \phi(-T_i, \hat{x}_i)$. After extracting suitable subsequences and re-indexing, we can assume T_i increasing, and $z_i \to z$ for some z with $|z| = \varepsilon$. According to (iii), there exists some $\tau > 0$ such that

$$|\phi(\tau, z)| \geq \eta > \varepsilon . \tag{2.31}$$

Let $w_i = \phi(\tau, z_i)$. Invoking again the uniform convergence of solutions on compact intervals, we should conclude that

$$w_i \to \phi(\tau, z) . \tag{2.32}$$

On the other hand, for large i, we have $T_i > \tau$, which implies by construction $|w_i| < \varepsilon$ for all i. This is a contradiction to (2.31), and the stability of the origin is proven.

Finally we show that the origin is attractive for (2.29). By virtue of stability, we can find an (open, nonempty) neighborhood U of the origin which is positively invariant with respect to (2.29) (see Corollary 1). Without restrictions, we can take $U \subset \mathcal{B}(\eta, 0)$, so that U is also bounded. For each $\hat{x} \in U$, the positive limit set of

---

[2] If $\eta \leq \varepsilon$, we can take $\delta < \eta/2$ and find $\hat{x}$ and T in such a way that $|\hat{x}| < \delta$ and $|\phi(-T, \hat{x})| \geq \varepsilon > \eta/2$, so that we can replace ε by $\eta/2$.

system[3] (2.29) $\Lambda^+(\hat{x})$ is nonempty, compact and contained in $\overline{U}$. Recall that $\Lambda^+(\hat{x})$ is invariant (positively and negatively). Hence, if there exists a point $y \in \Lambda^+(\hat{x})$, $y \neq 0$, then $\phi(t, y) \in \overline{U} \subset \overline{\mathcal{B}(\eta, 0)}$ for all $t \geq 0$, which contradicts the hypotheses. We conclude that $\Lambda^+(\hat{x}) = \{0\}$.   ∎

Remark 2.5 In the previous proof stability is an essential ingredient. We cannot prove that (ii) holds if we assume that the origin of (2.29) is simply attractive. As a counterexample, we can take an equilibrium point with a homoclinic orbit (see [8], Sect. 40).

The advantage of the properties expressed in (ii) and (iii) (Proposition 2.8) is that they involve only forward solutions. The formulation (iii) goes back to S. K. Persidskii (1931) and it is reported in the book [9] (Def. 2.6 p. 9), where the equilibria which satisfy (iii) are called *completely unstable*.

2.4.2 Discrete Time Case

The implications (i) $\Longrightarrow$ (ii) $\Longrightarrow$ (iii) of Proposition 2.8 hold, with some obvious modifications, also for invertible discrete dynamical systems. However, not surprisingly, for general (non-invertible) discrete dynamical systems the situation is more involved. We can propose several possible definitions of repulsivity.

Definition 2.12 We say that the fixed point x_* of system (2.1) is *strongly repulsive* if there exists an open neighborhood U of x_* with the following property: for each initial state $\hat{x} \in U$ ($\hat{x} \neq x_*$) there exists an integer $m > 0$ such that for each $n > m$ one has $x_n \notin U$, where $\{x_n\}$ is the solution of (2.1) initialized at $x_0 = \hat{x}$.

Definition 2.13 We say that the fixed point x_* of system (2.1) is *repulsive* if there exists an open neighborhood U of x_* with the following property: for each initial state $\hat{x} \in U$ ($\hat{x} \neq x_*$) and for each integer $m > 0$ there exists $n \geq m$ such that $x_n \notin U$, where $\{x_n\}$ is the solution of (2.1) initialized at $x_0 = \hat{x}$.

Definition 2.14 We say that the fixed point x_* of system (2.1) is *completely unstable* if there exists an open neighborhood U of x_* with the following property: for each initial state $\hat{x} \in U$ ($\hat{x} \neq x_*$) there exists n such that $x_n \notin U$ where $\{x_n\}$ is the solution of (2.1) initialized at $x_0 = \hat{x}$.

We stress that these definitions are applicable to non-invertible systems since they involve only forward solutions.

Definitions 2.12 and 2.14 correspond respectively to the discrete time versions of (ii) and (iii) of Proposition 2.8. Definition 2.13 is an intermediate property. Indeed, it is clear that strong repulsivity implies repulsivity, which in turn implies complete

---

[3] Equivalently, the negative limit set of system (2.28).

instability. Moreover, if x_* is completely unstable then it is trivially unstable. However, the notion of complete instability (and, a fortiori, repulsivity) should be not confused with instability. For instance, for the two-dimensional system of Example 1.3, the origin is unstable and not attractive, but it is not completely unstable (some trajectories are attracted).

The following examples illustrate how much nonlinear discrete dynamics can be counterintuitive, if your mental attitude is conformed to the study of continuous time dynamics.

Example 2.6 Let $f(x) : \mathbf{R} \to \mathbf{R}$ be defined by

$$f(x) = 1 - 2\left|x - \frac{1}{2}\right| = \begin{cases} 2x & \text{if } x \in \left(-\infty, \frac{1}{2}\right] \\ 2(1-x) & \text{if } x \in \left(\frac{1}{2}, +\infty\right). \end{cases}$$

The discrete system defined by $f(x)$ has two fixed points, the origin and $2/3$. It is not difficult to identify trajectories issuing from arbitrarily small neighborhoods of the origin, which converge to the origin (actually, such that $x_n = 0$ for large n), but also trajectories which converge to $2/3$. Clearly the origin is not stable. In fact, the origin is completely unstable. However, it is not attractive and not repulsive.

It is worthwhile noticing that the interval $[0, 1]$ is invariant for this system. The function f of this example, when restricted to $[0, 1]$, is the so-called *tent map*. It exhibits, as well known, a chaotic behavior (Fig. 2.3). ∎

Example 2.7 We now point out how to construct an example of a two-dimensional discrete system with a repulsive fixed point which is not strongly repulsive. In polar coordinates, we can take the pair of equations

$$\rho_{n+1} = \rho_n^q, \qquad \theta_{n+1} = 2\theta \bmod (2\pi)$$

with $q > 1$. Recall that the second equation defines a chaotic system on the unit circle. In particular, there are orbits which are dense on the unit circle. Looking at solutions which start in the proximity of the fixed point $\theta = 0$, $\rho = 1$, at the beginning we observe a diverging behavior. However, for $\hat{\rho} = 1$ and $\hat{\theta} \neq 0$, after a large number of iterations the solution may take again and repeatedly values in an arbitrarily small neighborhood of the fixed point. An apparently paradoxical behavior of solutions corresponding to $\hat{\rho} = 1$ and a small $\hat{\theta} < 0$ can be observed: while the system "rotates" counterclockwise, the points x_n gives the impression of moving in the clockwise sense (see Fig. 2.4). ∎

Fig. 2.3 Function f of
Example 2.6

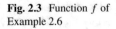

Fig. 2.3 Function f of
Example 2.6

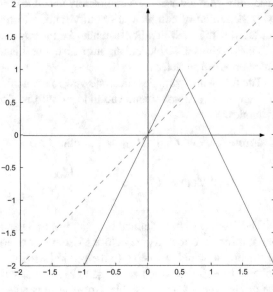

Fig. 2.4 Trajectories for
Example 2.7 with $q = 3/2$

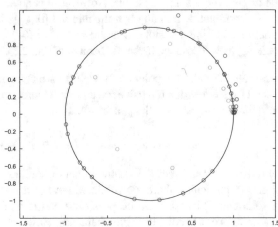

These examples point out also how the local behavior of a discrete system near a
fixed point can be actually affected by global features. More precisely, while complete instability has clearly a local nature, repulsivity cannot be ascertained without
investigating the global evolutions of trajectories.

2.4.3 Lyapunov-Like Criterion

Complete instability and repulsivity can be studied in the framework of Lyapunov method.

Theorem 2.9 *Let the system (2.1) with $f : \Omega \to \mathbf{R}^d$ be given, and let x_* be a fixed point. Assume that there exists a continuous function $V(x) : \Omega \to \mathbf{R}$ such that $V(x)$ and $V^+(x)$ are both positive definite at x_*. Then x_* is completely unstable.*

Proof First of all, we determine a number $\eta > 0$ such that

$$V(x) > 0 \quad \text{and} \quad V^+(x) > 0 \quad \text{for each } x \neq x_*, \ |x - x_*| < \eta \tag{2.33}$$

and $\mathcal{B}(\eta, x_*) \subset \Omega$. Let $G = \mathcal{B}(\eta/2, x_*)$, so that $\bar{G} \subset \Omega$.

We will prove the theorem by contradiction. If x_* is not completely unstable, then corresponding to η there exists a point $\hat{x} \in \mathcal{B}(\eta/4, x_*)$ ($\hat{x} \neq x_*$) such that $|x_n - x_*| < \eta/4$ for each integer $n \in \mathbf{N}$, where $\{x_n\}$ is the solution for which $x_0 = \hat{x}$. In particular, the solution $\{x_n\}$ is bounded and lies in G. Using (2.33) repeatedly, we get

$$0 < V(\hat{x}) = V(x_0) < V(x_1) < V(x_2) < \ldots \tag{2.34}$$

We are now in a position to apply Lemma 2.2. Since x_* is the unique point in G where $V^+(x) = 0$, we are led to conclude that $\lim_{n \to +\infty} x_n = x_*$ which yields, by continuity,

$$\lim_{n \to +\infty} V(x_n) = V(x_*) = 0 \tag{2.35}$$

which is impossible by (2.34). ∎

Theorem 2.10 *Let the system (2.1) with $f : \Omega \to \mathbf{R}^d$ be given, and let x_* be a fixed point. Assume that there exists a continuous function $V(x) : \Omega \to \mathbf{R}$ satisfying the following assumptions:*

(i) $V(x)$ is positive definite at x_*;
(ii) $V^+(x) = V(f(x)) - V(x) > 0$ for each $x \in \Omega, x \neq x_*$.

Then x_* is repulsive.

Proof According to (i), we can find a number $\eta > 0$ such that $\mathcal{B}(\eta, x_*) \subset \Omega$ and $V(x) > 0$ for each $x \neq x_*, |x - x_*| < \eta$. As in the previous theorem, we set $G = \mathcal{B}(\eta/2, x_*)$ and we argue by contradiction. If x_* is not repulsive, then there exist $\hat{x} \neq x_*$, with $|\hat{x} - x_*| < \varepsilon/4$, and an integer m such that for each $n > m$ one has $|x_n - x_*| < \varepsilon/4$, where $\{x_n\}$ is the solution for which $x_0 = \hat{x}$. This solution is bounded, and we have $x_n \in G$, possibly except for finitely many values of the index. The sequence of inequalities (2.34) holds also in this case, since (ii) is now assumed for each $x \in \Omega$

(note that (ii) implies in particular, by the induction argument, that $x_n \neq x_*$ for each n). We can therefore apply Lemma 2.2. A contradiction is obtained in the same way as in the previous proof.   ∎

The previous Example 2.6 points out that it is not sufficient to require *(ii)* in a neighborhood of x_*. The following theorem makes use of an alternative hypothesis.

Theorem 2.11 *Let the system (2.1) with $f : \Omega \to \mathbf{R}^d$ be given, and let x_* be a fixed point. Assume that there exists a continuous function $V(x) : \Omega \to \mathbf{R}$ satisfying the following assumptions:*

(i) $V(x)$ and $V^+(x)$ are both positive definite at x_*;
(ii) for each $x \neq x_*$, $x \in \Omega$, one has $f(x) \neq x_*$.

   Then x_* is repulsive.

Proof The proof can be carried on along the same lines as in the case of Theorem 2.10. The unique difference is that the sequence $V(x_n)$ in (2.34) begins to increase, in general, only starting from some index $\tilde{m}$. This fact, however, does not prevent us to apply Lemma 2.2.   ∎

Assumption (ii) of Theorem 2.11 is surely fulfilled if f is invertible.

2.5   Linear Systems

In this section we continue the study of the stability problem focusing on the special, but very important case of linear systems

$$x_{n+1} = Ax_n. \tag{2.36}$$

Without loss of generality, we can assume that the fixed point of interest is the origin. Indeed, if (2.36) has a fixed point $x_* \neq 0$, we may apply the transformation $y = x - x_*$. This is actually a topological equivalence, hence the stability properties of the fixed point are not changed. It is worth noticing that by virtue of linearity, even the form of the system remains unchanged:

$$y^+ = x^+ - x_* = Ax - x_* = Ax - Ax_* = A(x - x_*) = Ay .$$

Whatever we may say about the stability of the origin in terms of the matrix A, the same can be referred to any other fixed point $x_* \neq 0$.

2.5.1 Stability, Boundedness, Attraction

In the next statement we find convenient to use the Frobenius norm for matrices, but the result is obviously independent from choice of the matrix norm.

Theorem 2.12 *For the linear system (2.36), the following properties are equivalent.*

 (i) *The origin is a stable fixed point.*
 (ii) *All the solutions are bounded.*
 (iii) *The sequence $\{||A^n||_F\}$ is bounded.*

Proof First we prove that (i)$\Longrightarrow$(ii). According to the stability assumption, we fix a number $\delta > 0$ corresponding to $\varepsilon = 1$. For every $\hat{x} \in \mathbf{R}^d$ ($\hat{x} \neq 0$), let $K = \delta/(2|\hat{x}|)$ and $\hat{y} = K\hat{x}$. Clearly, $|\hat{y}| < \delta$, so that for each n one has $|A^n\hat{y}| < 1$. But $A^n\hat{y} = K A^n\hat{x}$, from which it follows $|A^n\hat{x}| < 1/K$ for each n.

Then we prove that (ii)$\Longrightarrow$(iii). Arguing by contradiction, we assume that the sequence $\{||A^n||_F\}$ or, equivalently, the sequence $\{||A^n||_F^2\}$, is not bounded. Denote by $a_{ij}^{[n]}$ the generic element of the matrix A^n, so that

$$||A^n||_F^2 = \sum_{ij} (a_{ij}^{[n]})^2 .$$

For at least one pair of indices i, j, the sequence $\{(a_{ij}^{[n]})^2\}$ must be unbounded. Without loss of generality, we may assume that this happens for $i = j = 1$. Now, take as initial state $\hat{x} = (1, 0, \ldots, 0)^{\mathbf{t}}$. The generated solution is

$$x_n = A^n\hat{x} = (a_{11}^{[n]}, a_{21}^{[n]}, \ldots, a_{d1}^{[n]})^{\mathbf{t}} .$$

Clearly, the sequence $\{|x_n|\}$ is unbounded, and this contradicts the assumption.

Finally we prove that (iii)$\Longrightarrow$(i). Let $M > 0$ such that $||A^n||_F < M$ for each n. Given $\varepsilon > 0$, let us take $\delta = \varepsilon/M$. Then for each $\hat{x}$ with $|\hat{x}| < \delta$ and each n we have

$$|A^n\hat{x}| \leq ||A^n||_F \cdot |\hat{x}| \leq M\delta = \varepsilon .$$

Hence, the origin is stable.   ∎

Now we focus our attention on the attraction property. First of all, we remark that if the origin is locally attractive for (2.36), then it is globally attractive, as well. This is a straightforward consequence of linearity. Moreover, if all the solutions converge to the origin, they are necessarily forward bounded. Hence for linear systems, by virtue of Theorem 2.12, (local) attraction implies stability.

Theorem 2.13 *For the linear system (2.36), the following properties are equivalent.*

 (i) *The origin is an attractive fixed point.*
 (ii) *For any matrix norm, $\lim_{n \to +\infty} ||A^n|| = 0$.*

Proof This time, for convenience, we agree that the chosen matrix norm is an induced norm. The proof that (ii)⟹(i) is straightforward. Indeed, for each initial state $\hat{x}$ and each n, we have

$$||A^n\hat{x}|| \le ||A^n|| \cdot |\hat{x}|$$

.

To prove the converse, we assume that for each $\hat{x}$, $|A^n\hat{x}| \to 0$. The following preliminary remark will be useful. As already noticed, if the origin is attractive then it is stable and hence the sequence $||A^n||$ is bounded. Let $\hat{x}_n$ be a sequence of initial states converging to a point z. We have

$$|A^n\hat{x}_n| \le |A^n z| + ||A^n|| \cdot |(\hat{x}_n - z)|$$

which yields $\lim_{n \to +\infty} |A^n\hat{x}_n| = 0$.

Now, assume by contradiction that there exists $\varepsilon > 0$ with the following property: for each integer i there exists an index $n_i > i$ such that $||A^{n_i}|| > \varepsilon$. Without loss of generality we can assume $n_i \to +\infty$. By definition of induced norm, for each i there exists a point $\hat{x}_i$ such that $|\hat{x}_i| = 1$ and $||A^{n_i}|| = |A^{n_i}\hat{x}_i|$. Taking possibly a subsequence and re-indexing, we may assume that for some z ($|z| = 1$), we have $\hat{x}_i \to z$. By construction, we conclude that $|A^{n_i}\hat{x}_i| > \varepsilon$ for each i while, by the above preliminary remark, we should have $|A^{n_i}\hat{x}_i| \to 0$. A contradiction. ∎

2.5.2   Stability, Attraction and Eigenvalues

To motivate the main achievements of this section, we consider first the particular case of system (2.36) under the assumption that $A = e^B$ for some square matrix B. In this case the solutions of (2.36) can be viewed as discretizations (with mesh 1) of the solutions of the differential system

$$\dot{x} = Bx \, . \tag{2.37}$$

It is well known that the origin is asymptotically stable for (2.37) if and only if all the eigenvalues of B have negative real part. On the other hand, every eigenvalue λ of e^B is of the form e^μ where μ is some eigenvalue of B. Having in mind Remark 2.3, we see that the origin is an asymptotically stable fixed point for (2.36) if and only if the modulus of all the eigenvalues of A is less than 1.

The main result of this subsection states that this last conclusion remains valid for any linear system of the form (2.36), even if the assumption $A = e^B$ is removed. Of course, a new proof is required. We start by a Lemma which establishes a link between the eigenvalues of the matrix A and the asymptotic behavior of the solutions of system

(2.36). The representation of the general solution expressed in Chap. 1, Sect. 1.3.3 is going to play here a crucial role. Recall in particular that the general solution of (2.36) can be written as a linear combination of the solutions of a fundamental set, each element of which being associated to an eigenvalue of A.

Lemma 2.3 *Let λ be an eigenvalue of the matrix A which defines system (2.36).*

(i) *If $|\lambda| < 1$ then all the solutions associated to λ converge to the origin for $n \to +\infty$.*

(ii) *If $|\lambda| > 1$ then all the nontrivial solutions associated to λ are unbounded.*

(iii) *If $|\lambda| = 1$ no nontrivial solution associated to λ converges to the origin.*

Sketch of the proof If λ is real with corresponding (real) eigenvalue v, then there is a solution of the form $x_n = \lambda^n v$. It is clear that this solution converges to the origin if and only if $|\lambda| < 1$. More precisely, if $|\lambda| > 1$ this solution is unbounded, while if $|\lambda| = 1$ it is constant or periodic of period 2.

If λ is complex (with nonzero imaginary part), also its conjugate $\overline{\lambda}$ is a (distinct) eigenvalue of A. Moreover, λ and $\overline{\lambda}$ have conjugate (linearly independent) eigenvectors $v, \overline{v}$. Let us consider the real solution $x_n = \lambda^n v + \overline{\lambda}^n \overline{v}$. Writing $\lambda = \rho(\cos\theta + i\sin\theta)$ and $v = u + iw$, we easily find

$$x_n = 2\rho^n[\cos(n\theta)u - \sin(n\theta)w]$$

(note that u, w are linearly independent in consequence of the linear independence of $v, \overline{v}$). The expression in the square bracket represents a sequence lying on a closed curve (actually, an ellipse) which does not cross the origin. As in the previous case, this solution converges to the origin if and only if $\rho = |\lambda| < 1$, it is unbounded if $\rho > 1$ and runs along the ellipse if $\rho = 1$.

If the construction of a solution associated to λ involves generalized eigenvalues the reasoning requires further considerations. The details are left to the reader. ∎

Remark 2.6 Some more remarks about the case (iii) of Lemma 2.3 are in order. If λ is of modulus 1 and if it is geometrically simple[4] then all the solutions associated to λ are bounded. If λ is not geometrically simple then the solutions associated to λ which involves generalized eigenvectors are unbounded. This happens for instance with the two-dimensional system defined by the matrix

$$A = \begin{pmatrix} 1 & 1 \\ 0 & 1 \end{pmatrix}.$$

∎

We now state the main result of this section. Recall that $\rho(A)$ denotes the spectral radius of A.

---

[4] This implies that no solution associated to λ requires the use of generalized eigenvectors.

Theorem 2.14 *The origin is an asymptotically stable fixed point for the system (2.36) if and only if $\rho(A) < 1$ i.e., for each eigenvalue λ of the matrix A one has $|\lambda| < 1$.*

Proof The statement is basically contained in Lemma 2.3, taking into account that the norms of the vectors v, u, w which appear in the proof can be taken as small as desired. As far as the sufficient part is concerned, there is an elegant alternative proof which makes use of the spectral norm $||A||_2$. Recall that

$$\lim_{n\to\infty} ||A^n||_2^{1/n} = \rho(A) .$$

If $\rho(A) < 1$ and ε is a positive number such that $\rho(A) + \varepsilon < 1$, we can find n_0 such that

$$\forall n > n_0, \quad ||A^n||_2 \le (\rho(A) + \varepsilon)^n . \tag{2.38}$$

Hence, for each initial state $\hat{x}$,

$$||A^n \hat{x}|| \le ||A^n|| \cdot |\hat{x}| \le (\rho(A) + \varepsilon)^n |\hat{x}|$$

from which the conclusion follows.   ∎

Theorem 2.15 *The origin is a stable fixed point for the system (2.36) if and only if for each eigenvalue λ of the matrix A we have $|\lambda| \le 1$ and, in addition, all the eigenvalues λ for which $|\lambda| = 1$ (if any) are geometrically simple.*

We emphasize that the conditions of Theorems 2.14 and 2.15 must hold for all (both real and complex) eigenvalues of A. Alternative proofs of Theorems 2.14 and 2.15 based on the Jordan canonical form are given in [2].

2.5.3   Lyapunov Functions Approach

When dealing with linear systems, the search for possible Lyapunov functions can be limited to quadratic functions of the form $V(x) = x^t P x$, where P is a symmetric, positive definite real matrix. The function $V^+(x)$ associated to such $V(x)$ and system (2.36) is $V^+(x) = x^t(A^t P A - P)x$. In what follows, a crucial role is played by the matrix equation

$$A^t X A - X = -Q \tag{2.39}$$

where the square matrix Q is given, and X is the unknown. It is called the *discrete matrix Lyapunov equation*.

Remark 2.7 The operator $X \mapsto \mathcal{L}(X) = A^t X A - X$ is linear on the vector space of the square $d \times d$ matrices. Thus, (2.39) can be viewed as a non-homogeneous algebraic system. In fact, (2.39) can be rewritten

$$(A^t \otimes A^t)\text{vec } X - \text{vec } X = -\text{vec } Q$$

where $\otimes$ represents the Kronecker product of matrices and the symbol "vec" denotes the vectorization operator. ∎

From this last remark and the properties of the Kronecker product, recalling that A and A^t have the same eigenvalues, we readily get the following result.

Lemma 2.4 *A number $r \in \mathbf{C}$ is an eigenvalue of $\mathcal{L}$ if and only if there exists a pair λ, μ of (not necessarily distinct) eigenvalues of A such that $r = \lambda\mu - 1$.*

Consequently, the equation (2.39) has a unique solution P for any given matrix Q if and only if for each pair λ, μ of eigenvalues of A one has $\lambda\mu \neq 1$.

Note that if (2.39) has a unique solution P and if Q is symmetric, then P is symmetric, as well. Indeed, if $A^t P A - P = -Q$ then also $A^t P^t A - P^t = -Q$ and hence, by uniqueness, $P = P^t$.

Next theorem is the main result of this subsection.

Theorem 2.16 *Let the system (2.36) be given. The following statements are equivalent.*

(i) The origin is an asymptotically stable fixed point.
(ii) The matrix equation (2.39) with $Q = I$ admits a unique symmetric, positive definite real solution P.
(iii) For each symmetric, positive definite real matrix Q the matrix equation (2.39) admits a unique symmetric, positive definite real solution P.

Proof Assume that (i) is true. By Theorem 2.14, the condition required by Lemma 2.4 is fulfilled. Hence, Eq. (2.39) must have a solution P for each given matrix Q, and this solution is unique. Now we show how such P can be explicitly represented.

Consider first the case $Q = I$. We can multiply both members of the identity $A^t P A - P = -I$ by $(A^t)^k$ on the left and by A^k on the right, which gives

$$(A^t)^{k+1} P A^{k+1} - (A^t)^k P A^k = -(A^t)^k A^k.$$

This yields

$$\sum_{k=0}^{n} \left[(A^t)^{k+1} P A^{k+1} - (A^t)^k P A^k \right] = (A^t)^{n+1} P A^{n+1} - P = -\sum_{k=0}^{n} (A^t)^k A^k.$$

Finally taking the limit for $n \to \infty$, we obtain $P = \sum_{k=0}^{\infty} (A^t)^k A^k$. As expected, P is symmetric. Moreover, for each n, $x^t (A^t)^n A^n x = x^t (A^n)^t A^n x = (A^n x)^t A^n x =$

$||A^n x||_F^2 \geq 0$, and if $x \neq 0$ the inequality is strict at least for $n = 0$. Hence P is positive definite. We have so completed the proof that (i) implies (ii).

Next we give a direct proof that (ii) implies (iii). Since Q is symmetric and positive definite, it admits the factorization $Q = R^t R$ where R is nonsingular [10]. Let $\tilde{A} = RAR^{-1}$. The matrices A and $\tilde{A}$ have the same eigenvalues, and so for the first part of this proof there exists a unique symmetric positive definite matrix $\tilde{P}$ such that

$$\tilde{A}^t \tilde{P} \tilde{A} - \tilde{P} = -I$$

or

$$(RAR^{-1})^t \tilde{P} RAR^{-1} - \tilde{P} = -I.$$

Multiplying by R^t on the left and by R on the right and setting $P = R^t P R$, we find

$$A^t P A - P = -Q$$

as desired. We also can see that $P = \sum_{k=0}^{\infty} (A^t)^k Q A^k$.

Finally, the proof that (iii) implies (i) is a straightforward application of Theorem 2.4, with $V(x) = x^t P x$. ∎

The previous theorem enlightens in particular the flexibility of the matrix equation (2.39). Roughly speaking, it states that in order to ascertain the asymptotic stability of system (2.36) it is sufficient to find a pair of symmetric, positive definite real matrices P, Q such that (2.39) holds. Of course, the choice $Q = I$ is often convenient to simplify computations. We may also say that the origin is asymptotically stable for system (2.36) if and only if the matrix $A^t P A - P$ is negative definite for some symmetric definite positive P. Other results about matrix equations related to (2.39) can be found in [3].

The following theorem provides a characterization of stability.

Theorem 2.17 *Let the system (2.36) be given. The origin is a stable fixed point if and only if there exists a symmetric, positive definite real matrix P such that the matrix $A^t P A - P$ is positive semidefinite.*

The "if" part of the previous theorem is a particular case of Theorem 2.1. We do not give the proof of the converse statement, which can be found for instance in [1, 3, 11]. Instead, for later use, we are more interested in the following statement.

Proposition 2.9 *Let the system (2.36) be given, and assume that there is an eigenvalue of the matrix A whose modulus is strictly greater than 1 i.e., $\rho(A) > 1$. Assume also that for each pair of eigenvalues λ, μ of A one has $\lambda\mu \neq 1$, and let P be the unique symmetric solution of the equation (2.39) with $Q = I$. Then, there exist points $x \neq 0$ for which $x^t P x < 0$.*

Proof Existence and uniqueness of P is guaranteed by Lemma 2.4. Moreover, it is obvious that $P \neq 0$. Clearly, P cannot be positive definite; otherwise, by Theorem 2.16 the origin would be asymptotically stable, which is impossible by Theorem 2.14. In fact, P is not even positive semidefinite. Indeed, if P is positive semidefinite and if $x^t P x = 0$ for some $x \neq 0$, using (2.39) we obtain

$$x^t A^t P A x = y^t P y = -x^t x$$

where we set $y = Ax$. We should therefore have that $y^t P y \geq 0$, while $-x^t x < 0$. The statement is so proven by exclusion. ∎

2.5.4 Repulsivity

Concerning repulsivity of linear systems, we have the following preliminary result.

Lemma 2.5 *If the origin is completely unstable for the system (2.36), the matrix A is nonsingular. In fact, for each eigenvalue λ of the matrix A one has $|\lambda| > 1$.*

Proof Assume that $\lambda = 0$ is an eigenvalue of A. Then we could find solutions of the form $x_0 = \hat{x}, x_1 = x_2 = \ldots = 0$ with nonzero $\hat{x}$ of arbitrarily small norm. Hence, the origin would be not completely unstable. To prove the second statement (which is basically included in Lemma 2.3), we assume first that A admits some eigenvalue λ such that $|\lambda| < 1$. Then, there are solutions $\{x_n\}$ converging to the origin and hence bounded. Say $|x_n| < M$ for each n. Then for each $\eta > 0$ the solution $y_n = \eta x_n / M$ satisfies $|y_n| < \eta$ for each n. This means that the origin is not completely unstable. Finally if there is some eigenvalue λ such that $|\lambda| = 1$, we will have again bounded solutions, and we can repeat the previous argument. ∎

Actually, repulsive linear systems can be characterized in terms of eigenvalues.

Theorem 2.18 *The following facts are equivalent.*

(i) The origin is strongly repulsive for the system (2.36).
(ii) The origin is completely unstable for the system (2.36).
(iii) For each eigenvalue λ of the matrix A one has $|\lambda| > 1$.
(iv) A is nonsingular and the origin is an asymptotically stable fixed point for the system $x^+ = A^{-1}x$.

Proof The implication (i) $\Longrightarrow$ (ii) is consequence of the definitions. The implication (ii) $\Longrightarrow$ (iii) reduces to Lemma 2.5. Assume that (iii) is true. Obviously, the matrix A is nonsingular. Moreover, we know that the eigenvalues of A^{-1} have the form $\mu = 1/\lambda$, where λ is some eigenvalue of A. Hence, for each eigenvalue μ of A^{-1} we have $|\mu| < 1$, and (iv) is true, as well. Finally, to prove that (iv) $\Longrightarrow$ (i), we reverse the previous argument. If A is nonsingular then A^{-1} exists and if all its eigenvalues satisfy $|\mu| < 1$, then all the eigenvalues of A satisfy $|\lambda| > 1$. The conclusion follows taking into account the form of the solutions (Lemma 2.3) and Sect. 1.3.3. ∎

Remark 2.8 According to Theorem 2.18, from now on we use simply the term "repulsive" to denote a linear system satisfying one of the above properties (i), (ii), (iii) or (iv).   ■

Next we give a characterization of repulsive linear systems in the framework of Lyapunov method.

Theorem 2.19 *The system (2.36) is repulsive if and only if there exist symmetric, positive definite real matrices P, R such that*

$$A^t P A - P = R . \tag{2.40}$$

Proof Assume that (2.36) is repulsive. Then, by Theorem 2.18 the system defined by A^{-1} is asymptotically stable at the origin. Therefore, according to Theorem 2.16 there exists some symmetric, positive definite real matrix P, Q such that

$$(A^{-1})^t P A^{-1} - P = -Q . \tag{2.41}$$

Multiplying by A^t on the left and by A on the right, we get

$$A^t P A - P = R \tag{2.42}$$

with $R = A^t Q A$. Vice-versa, if (2.42) holds, then A must be nonsingular. Indeed, in the opposite case we should have

$$\det (A^t P A) = (\det A)^2 \det (P) = 0 = \det (P + R)$$

which is impossible since P and R are positive definite. Hence we can repeat the previous computation multiplying by $(A^{-1})^t$ and A^{-1}, recovering (2.41). The conclusion follows using again Theorems 2.16 and 2.18.   ■

2.6 Linearization

One reason of interest in linear systems theory is that it provides information about the local behavior of the solutions of certain nonlinear systems around fixed or periodic points, by means of a procedure called *linearization*. Beyond the assumptions established at the beginning of this chapter, in this section we need to require that f is differentiable.

2.6.1   *Fixed Points*

One typical result in this direction, is the following theorem.

Theorem 2.20 *(Stability by first approximation) Consider a discrete dynamical system (2.1) and assume that $f \in C^1(\Omega)$. Assume further that it has a fixed point $x_* \in \Omega$.*

If for all the eigenvalues λ of the matrix $A = (Df)(x_*)$ we have $|\lambda| < 1$, then x_* is locally asymptotically stable for (2.1).

If there exists an eigenvalue λ of the matrix $A = (Df)(x_*)$ such that $|\lambda| > 1$, then x_* is unstable.

If for all the eigenvalues λ of the matrix $A = (Df)(x_*)$ we have $|\lambda| > 1$, then x_* is completely unstable.

Roughly speaking, this theorem states that the local stability properties of (2.1) at x_* can be related to the stability properties of the linear system

$$x^+ = Ax \qquad\qquad (2.43)$$

at the origin, where $A = (Df)(x_*)$. System (2.43) is called the *first order approximation* or the *linearization* of (2.1).

Proof of Theorem 2.20 For simplicity, and without loss of generality, we will assume that $x_* = 0$. By applying the first order Taylor formula we have

$$x_{n+1} = f(x_n) = f(0) + (Df)(0)(x_n) + g(x_n) = Ax_n + g(x_n)$$

where $g(x) = o(x)$ for $x \to 0$ that is, $\lim_{x\to 0} |g(x)|/|x| = 0$.

We prove the first statement. Since by assumption system (2.43) is asymptotically stable at the origin, according to Theorem 2.16 it admits a Lyapunov function of the form $V(x) = x^t Px$ such that $A^t PA - P = -I$.

We are going to apply the same Lyapunov function to system (2.1). After some computations, we get

$$V^+(x) = -x^t x + 2x^t A^t Pg(x) + (g(x))^t Pg(x)$$
$$= -x^t x \left[1 + 2\frac{x^t A^t Pg(x)}{|x|^2} + \frac{(g(x))^t Pg(x)}{|x|^2} \right].$$

Since $g(x) = o(x)$, it is straightforward to conclude that there exists a number $r > 0$ such that

$$V^+(x) < -\frac{1}{2}x^t x$$

for $x \in \mathcal{B}(r, 0)$, $x \neq 0$. Hence $V^+(x)$ is negative definite at $x_* = 0$ and, by Theorem 2.4, the origin is asymptotically stable for system (2.1).

As far as the second statement is concerned, we limit ourselves to the particular (but generic) case where for each pair λ, μ of eigenvalues of A we have $\lambda\mu \neq 1$. According to Proposition 2.9, there is matrix P for which $A^t P A - P = -I$ and the function $V(x) = x^t P x$ takes a negative value for some $x \neq 0$ of arbitrarily small norm. Applying the same function $V(x)$ to system (2.1) and repeating the same arguments as before, we see that $V^+(x)$ is negative definite at $x_* = 0$. Thus, we may invoke Theorem 2.7 to get the conclusion. To get rid of the additional assumption about the eigenvalues of A one more subtle argument is needed: we refer the reader to [3], p. 38.

To show the last statement, we construct a quadratic function $V(x) = x^t P x$ for (2.43) on the base of Theorems 2.18 and 2.19, and apply it to the given system. Using the usual approximation argument we get the conclusion by virtue of Theorem 2.9,   ∎

It is important to emphasize the local nature of Theorem 2.20. In particular, in case the asymptotic stability of system (2.1) is ascertained only on the base of Theorem 2.20, there is no hope to obtain information about the size of the set of attraction. Moreover, we cannot achieve conclusions about forms of repulsivity stronger than complete instability, or in order to exclude attractivity.

Example 2.8 Consider again the Verhulst equation in rescaled form (compare with Examples 1.11, 1.14, 2.5)

$$x^+ = f(x) = a(1 - x)x \tag{2.44}$$

as a dynamical system on the whole of $\mathbf{R}$. If $a \notin \{0, 1\}$, it admits two distinct fixed points: the origin and $x_* = (a - 1)/a$. Since $f'(x) = a - 2ax$, the Theorem of stability by first approximation provides the following information: the origin is locally asymptotically stable for $|a| < 1$, and becomes completely unstable for $|a| > 1$. The fixed point $x_* = (a - 1)/a$ is locally asymptotically stable if $1 < a < 3$, and completely unstable for $a < 1$ and $a > 3$.

Notices also that for every $a > 1$, the nontrivial fixed point x_* belongs to the interval of biological interest $[0, 1]$.   ∎

Example 2.9 Consider the Lotka-Volterra model on the whole of $\mathbf{R}^2$

$$\begin{cases} x_{n+1} = ax_n - px_n y_n \\ y_{n+1} = by_n + px_n y_n \end{cases} \tag{2.45}$$

with $a > 1$, $0 < b < 1$, $p \geq 0$. We already know that this system admits two fixed points: the origin and the point $((1 - b)/p, (a - 1)/p)$. We have

$$(Df)(x, y) = \begin{pmatrix} a - py & -px \\ py & b + px \end{pmatrix}.$$

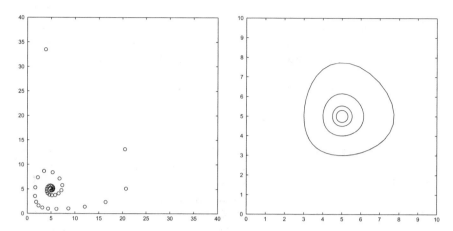

Fig. 2.5 Orbits of the Lotka-Volterra model: discrete case on the left ($a = 3/2, b = 1/2, p = 1/10$), continuous case on the right ($a = 1/2, b = 1/2, p = 1/10$). In the discrete case, for large n the solution leaves the region of biological significance

In particular, the eigenvalues of the matrix $(Df)(0, 0)$ are a and b. According to our assumptions, one of them is greater than 1: the origin is unstable. As far as the second fixed point is concerned, we have

$$(Df)((1 - b)/p, (a - 1)/p) = \begin{pmatrix} 1 & b - 1 \\ a - 1 & 1 \end{pmatrix}$$

and since $(a - 1)(b - 1) < 0$, its eigenvalues are complex conjugate:

$$\lambda = 1 \pm i\sqrt{(a - 1)(1 - b)} \, .$$

Clearly, $|\lambda| > 1$ and the fixed point is completely unstable. It is worth noticing that this conclusion is in contrast to what happens in the corresponding continuous time model, where the nontrivial equilibrium can be proved to be stable (not asymptotically). The simulations reported in Fig. 2.5 illustrate the different behaviors.

This inconsistency possibly points out that the proportional increment hypothesis is not satisfied for this application. ∎

It is important to notice that the statements of Theorem 2.20 are not necessary and sufficient conditions. The case where A possesses some eigenvalue of modulus 1 is not covered by the theorem and it is indeed possible to construct examples showing that in this case, the stability of x_* actually depends on the nonlinear terms.

Example 2.10 Consider in $\mathbf{R}$ the discrete dynamical system

$$x^+ = f(x) = x + x^3 \, .$$

The unique fixed point is the origin $x = 0$, and $f'(0) = 1$. If $x > 0$, we clearly have $x^+ = f(x) > x$. Any solution corresponding to a positive initial state is an increasing sequence, and hence it cannot converge to the fixed point.   ∎

Example 2.11 Now we consider in **R** the discrete dynamical system

$$x^+ = f(x) = x - x^3 \, .$$

Again, the unique fixed point is the origin $x = 0$, and $f'(0) = 1$. However, this time the origin is locally asymptotically stable. This can be proven on the base of Theorem 2.4. As a Lyapunov function, we can use, for instance, $V(x) = x^2$. It is easily seen that the conditions of Theorem 2.4 are satisfied for each $x \in (-\sqrt{2}, \sqrt{2})$. We may also take $V(x) = |x|$, and we get the same conclusion. It seems reasonable to conjecture that $\mathcal{A}(0) = (-\sqrt{2}, \sqrt{2})$: this can be actually confirmed by noticing that the points $x = \pm\sqrt{2}$ are periodic for our system.   ∎

Definition 2.15 If the matrix $A = (Df)(x_*)$ has at least one eigenvalue of modulus 1, we say that x_* is a *critical* fixed point of (2.1). Instead, if for each eigenvalue λ of the matrix $A = (Df)(x_*)$ one has $|\lambda| \neq 1$, we say that x_* is a *hyperbolic* fixed point of (2.1).

When x_* is hyperbolic, we may also say the (2.1) is *linearizable* at x_*. Two particular typologies of hyperbolic fixed points have been considered in Theorem 2.20: the case where all the eigenvalues of the matrix $A = (Df)(x_*)$ are inside the unit open disc of **C** and the case where all the eigenvalues of the matrix $A = (Df)(x_*)$ are outside the unit closed disc of **C**; in more qualitative terms, in the former case x_* is sometimes qualified as a *sink*, in the latter case as a *source*. When $d = 2$ and the eigenvalues λ_1, λ_2 of the matrix $A = (Df)(x_*)$ are such that $|\lambda_1| < 1 < |\lambda_2|$ (as it happens for the origin in Examples 1.3 and 2.9, see Fig. 2.6) the fixed point is hyperbolic: it is unstable, but not completely unstable. In such a situation, x_* is qualified as a *saddle point*.

Example 2.12 In the case of the discrete SIR model

$$\begin{cases} x_{n+1} = x_n - p x_n y_n \\ y_{n+1} = y_n - s y_n + p x_n y_n \end{cases} \tag{2.46}$$

($x \geq 0$, $y \geq 0$, $0 < b = 1 - s < 1$, $p \geq 0$), there are infinitely many fixed points. They are precisely the points of the form $(x, 0)$. We have

$$(Df)(x, 0) = \begin{pmatrix} 1 & -px \\ 0 & b + px \end{pmatrix} .$$

All of them are critical, because of the eigenvalue 1. Moreover, when $x = (1 - b)/p$ the critical eigenvalue is multiple. Recall that at $x = (1 - b)/p$ the monotonicity of the first component of the solutions changes. Moreover, for $x < (1 - b)/p$

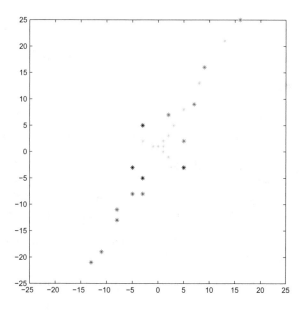

Fig. 2.6 Trajectories of the system of Example 1.3 around the saddle point

the absolute value of the eigenvalue is less then 1, while it is greater than 1 for $x > (1 - b)/p$. The slope of the corresponding eigenvector changes accordingly. All these remarks are consistent with the behavior of the trajectories, as shown by Fig. 1.11. ∎

2.6.2 Periodic Points

The theorem of stability by first approximation (Theorem 2.20) extends to periodic points, by virtue of Proposition 2.5.

Corollary 3 *Consider a discrete dynamical system (2.1) and assume that $f \in C^1(\Omega)$. Let p be a periodic point of period k, and let Γ be the cycle containing p. If for all the eigenvalues λ of the matrix $(Df^{[k]})(p)$ we have $|\lambda| < 1$, then Γ is locally asymptotically stable for system (2.1). If there exists an eigenvalue λ of the matrix $(Df^{[k]})(p)$ such that $|\lambda| > 1$, then Γ is unstable.*

As far as the previous statement is concerned, it is remarkable that the evaluation of the derivative $(Df^{[k]})$ can be done at any point $q \in \Gamma$. To motivate this claim, consider for simplicity the case $k = 2$. Let p_0 be a fixed point of the map $f^{[2]}$ which is not a fixed point of f. Namely, let

$$p_1 = f(p_0) \neq p_0 \quad \text{and} \quad p_0 = f(f(p_0))$$

and so also $p_0 = f(p_1)$ and $p_1 = f(f(p_1))$. According to the chain rule, we have

$$D(f \circ f)(p_0) = (Df)(f(p_0)) \cdot (Df)(p_0) \tag{2.47}$$

while

$$D(f \circ f)(p_1) = (Df)(f(p_1)) \cdot (Df)(p_1) = (Df)(p_0) \cdot (Df)(f(p_0)). \tag{2.48}$$

Comparing (2.47) and (2.48), we notice that $D(f \circ f)(p_0)$ and $D(f \circ f)(p_1)$ are obtained as a product of the same pair of factors, but exchanging the order. Thus $D(f \circ f)(p_0)$ and $D(f \circ f)(p_1)$ are not necessarily equal, but nevertheless they have the same eigenvalues, as pointed out by the following lemma.

Lemma 2.6 *Let M, N square matrices of the same size. Then MN and NM have the same eigenvalues.*

Proof Let $\lambda \neq 0$ be an eigenvalue of MN. Then for some $v \neq 0$, we may write

$$MNv = \lambda v \tag{2.49}$$

from which $NMNv = \lambda Nv$ or $NMu = \lambda u$ where $u = Nv$. We must have $u \neq 0$; otherwise, we will find a contradiction in (2.49). Thus we see that λ is also an eigenvalue of NM.

If MN has an eigenvalue $\lambda = 0$, we must have

$$0 = \det(MN) = \det M \cdot \det N = \det(NM)$$

which shows that $\lambda = 0$ is also an eigenvalue of NM. We complete the proof by exchanging the roles of M and N. ∎

Example 2.13 We already remarked that the system in Example 2.11 has a periodic point at $x = \pm\sqrt{2}$. The previous corollary can be applied to show that the corresponding cycle is unstable. ∎

Example 2.14 Looking for examples of asymptotically stable cycles, we may consider the system

$$x^+ = \frac{-2x}{1 + |x|} . \tag{2.50}$$

The origin is the unique fixed point and, by first approximation, we see that it is completely unstable. The second-iteration system is

$$x^+ = \frac{4x}{1 + 3|x|}$$

Fig. 2.7 Graph of the second iterate of the logistic map for $a = 3.1$ (Example 2.15)

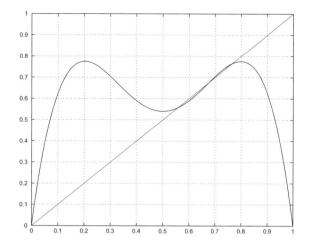

which possesses three fixed points, $x = 0$, $x = \pm 1$. The set $\{-1, 1\}$ is a periodic orbit of period 2 for (2.50). Using the version of the first approximation method for cycles, we see that this orbit is asymptotically stable. ∎

Example 2.15 We already found in Example 1.11 that the logistic equation (2.44)

$$x^+ = f(x) = ax(1 - x)$$

has a cycle of period 2 for $a = 4$. Actually, it is not difficult to check that a cycle of period 2 exists for any $a > 3$. Using Corollary 3, it is possible to check that this cycle is asymptotically stable for $3 < a < 1 + \sqrt{6} = 3.4495\ldots$. Figures 2.7 and 2.8 show the graph and some solutions of the second iterate of the logistic equation with $a = 3.1$. Let x_* be the (nontrivial) fixed point which is also a fixed point of (2.44) and let p, q ($p < x_* < q$) be the new fixed points (periodic points for (2.44)). Let moreover a, b, c be the three points such that $f(a) = f(b) = f(c) = x_*$, $0 < a < b < p < x_* < q < c < 1$. It may be easily conjectured (and confirmed by numerical simulations) that the set of attraction of p (as a fixed point of $f^{[2]}$) is $(0, a) \cup (b, x_*) \cup (c, 1)$ while the set of attraction of p (as a fixed point of $f^{[2]}$) is $(a, b) \cup (x_*, c)$. The set of attraction of the cycle is the union of all these subintervals, that is $[0, 1] \setminus \{0, a, b, x_*, c, 1\}$. All these sets are open but not connected. ∎

Example 2.16 In this example we consider a variant of the discrete logistic equation. In (2.44), the term $a(1 - x)$ can be thought of as a state dependent incremental factor, regulated by the intraspecific competition effect: when x increases, the incremental factor decreases. Now we want to consider a slightly different model

$$x_{n+2} = a(1 - x_n)x_{n+1} \tag{2.51}$$

Fig. 2.8 Graphs (n vs. x) of some solutions of the second iterate of the logistic map for $a = 3.1$ (Example 2.15)

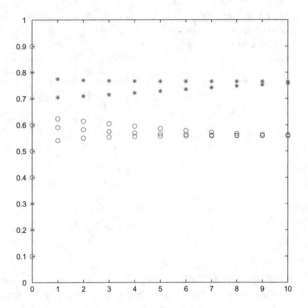

where at each step, the change of the incremental factor arises in a similar way, but with a one time unit of delay. The second order discrete dynamical system (2.51) can be rewritten as a system in $\mathbf{R}^2$:

$$\begin{cases} x_{n+1} = ax_n(1 - y_n) \\ y_{n+1} = x_n \end{cases} \tag{2.52}$$

To guarantee the biological significance, we should also require that x (and so also y) belongs to the interval $[0, 1]$. However, the square $Q = [0, 1] \times [0, 1]$ is not positively invariant with respect to (2.52).

Claim If $1 < a < 2$, the largest positively invariant subset of Q is[5]

$$M = Q \cap \left\{ (x, y) : y \geq 1 - \frac{1}{ax} \right\}.$$

Proof First of all, we remark that from $1 < a < 2$ it follows $0 < 1 - \frac{1}{a} < \frac{1}{a}$. Moreover, it is easy to check the following facts.

Fact A. If $(x_n, y_n) \in Q$ then $x_{n+1} \geq 0$ and $y_{n+1} \in [0, 1]$. If in addition $y_n \geq 1 - 1/ax_n$ then we also have $x_{n+1} \leq 1$.

Fact B. If $y_n \geq 1 - 1/a$ then $x_{n+1} \leq x_n$.

Fact C. If $y_n \leq 1 - 1/a$ then $x_{n+1} \geq x_n$.

---

[5] The curve $y = 1 - 1/ax$ is a hyperbola which intersects the x-axis at $x = 1/a$.

In particular, Fact A implies that $(x_{n+1}, y_{n+1}) \in Q$ whenever $(x_n, y_n) \in M$. Moreover, according to the previous observations, it is not difficult to check that actually $(x_{n+1}, y_{n+1}) \in M$ whenever $(x_n, y_n) \in M$ in the following cases.
Case 1. $1 - 1/a \le x_n \le 1$, $y_n \in [0, 1]$;
Case 2. $0 \le x_n \le 1 - 1/a$, $1 - 1/a \le y_n \le 1$.
Thus it remains to consider the case:
Case 3. $0 \le x_n \le 1 - 1/a$, $0 \le y_n \le 1 - \frac{1}{a}$.
We actually show that if (x_n, y_n) satisfies the conditions of Case 3, then

$$x_{n+1} \le t_n \quad \text{where} \quad t_n = \frac{1}{a(1 - x_n)} . \tag{2.53}$$

The inequality (2.53) implies our statement. Indeed,

$$t_n = \frac{1}{a(1 - x_n)} \iff x_n = 1 - \frac{1}{at_n}$$

and $y_{n+1} = x_n$. Hence, if $x_{n+1} \le t_n$ then automatically

$$y_{n+1} \ge 1 - \frac{1}{ax_{n+1}} .$$

To prove (2.53), we argue by contradiction. Assume that for some choice of (x_n, y_n) satisfying the conditions of Case 3, we find $x_{n+1} > t_n$, that can be rewritten as

$$ax_n(1 - y_n) > \frac{1}{a(1 - x_n)} .$$

Since $1 - y_n \le 1$ and $1 - x_n > 0$, from this last inequality we infer

$$x_n(1 - x_n) > \frac{1}{a^2}$$

or, equivalently,

$$x_n^2 - x_n + \frac{1}{a^2} < 0 .$$

But, the left-hand side (as a function of x_n) has a minimum for $x_n = 1/2$ and the value of the minimum is positive (a being less than 2). We have so found a contradiction. $\blacksquare$

In fact, it is also possible to prove that if $1 < a < (1 + \sqrt{5})/2$ and (x_n, y_n) satisfies the conditions of Case 3 above, then $x_{n+1} \leq 1/a$.

System (2.52) admits two fixed points: the origin and, if $a \neq 0$,

$$P_a = \left(1 - \frac{1}{a}, 1 - \frac{1}{a}\right).$$

From now on we limit ourselves to the case $a > 1$, so that in particular the fixed points are distinct and $P_a \in M$. Next we examine their stability using the first approximation. The eigenvalues of the linearized system at the origin are a and zero, and since $a > 1$ the origin is a saddle point. If $1 < a \leq 5/4$, the eigenvalues of the linearized system at P_a are real; more precisely, we have

$$\lambda_{1,2} = \frac{1 \pm \sqrt{5 - 4a}}{2}$$

and it is not difficult to check that they are both positive and less than 1. Hence, the point P_a is locally asymptotically stable. For $a > 5/4$ the eigenvalues are:

$$\lambda_{1,2} = \frac{1 \pm i\sqrt{4a - 5}}{2}.$$

An easy computation shows that $|\lambda_{1,2}|^2 = a - 1$, so that $|\lambda_{1,2}| < 1$ as far as $a < 2$, and P_a is still asymptotically stable.

For completeness, we finally remark that when $a = 2$ the point P_a is critical, since $|\lambda_{1,2}| = 1$. When a becomes greater than 2, we have $|\lambda_{1,2}| > 1$ and the point P_a is no more stable.

It is interesting to compare the qualitative behavior of the trajectories of (2.52) with those of its continuous time version deduced under the proportional increment assumption

$$\begin{cases} \dot{x} = (a - 1)x - axy \\ \dot{y} = x - y. \end{cases} \tag{2.54}$$

The equilibria of (2.54) are again the origin and $P_a = (1 - 1/a, 1 - 1/a)$. As far as $1 < a < 2$, the dynamical behavior of the trajectories of (2.52) around P_a agrees with that of the trajectories of (2.54) (see Fig. 2.9 and 2.10). But, for $a > 2$, the behaviors crucially differ. Indeed, P_a is asymptotically stable for (2.54) for every $a > 1$ while, as already noticed, P_a becomes unstable for (2.52) when $a > 2$.   ∎

Example 2.17 In the literature about population theory (see for instance [12]) the following variant of the Lotka-Volterra model has been studied:

$$\begin{cases} x_{n+1} = ax_n - \gamma x_n^2 - px_n y_n \\ y_{n+1} = by_n + px_n y_n. \end{cases} \tag{2.55}$$

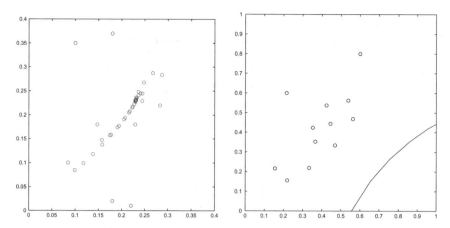

Fig. 2.9 On the left: trajectories of system (2.52) with $a = 1.3$ corresponding to the initial states $(0.22, 0.01)$, $(0.18, 0.02)$, $(0.18, 0.38)$, $(0.1, 0.35)$. On the right: the trajectory of system (2.52) with $a = 1.8$ corresponding to the initial state $(0.6, 0.8)$

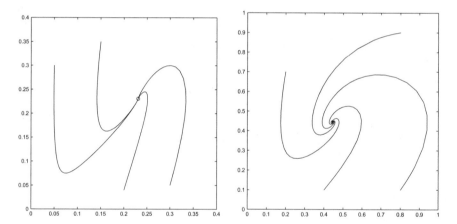

Fig. 2.10 On the left: trajectories of system (2.54) with $a = 1.3$. On the right: trajectories of system (2.54) with $a = 1.8$. In the second case the trajectories exhibit an oscillatory behavior

It includes, for the prey population, the intraspecific competition effect with coefficient $\gamma > 0$ borrowed from the logistic equation. As in the classical Lotka-Volterra model, $a > 1$, $0 < b < 1$ and $p \geq 0$. After renormalization, it is possible to put the model in the form

$$\begin{cases} x_{n+1} = ax_n(1 - x_n) - x_n y_n \\ y_{n+1} = by_n + qx_n y_n \end{cases} \tag{2.56}$$

where $q = ap/\gamma$. The region of biological interest is the quadrant $Q = \{(x, y) \in \mathbf{R}^2 : x \geq 0, y \geq 0\}$ although mathematically, (2.56) makes sense for each $(x, y) \in \mathbf{R}^2$. However, Q is not positively invariant for (2.56): for instance, the righthand-side of the first equation in (2.56) takes the value -1 when $x = y = 1$.

The fixed points of (2.56) are the origin of $\mathbf{R}^2$, the point

$$P_1 = \left(\frac{a-1}{a}, 0\right)$$

(extinction of predators) and, if $q \neq 0$,

$$P_2 = \left(\frac{1-b}{q}, \frac{q(a-1) + a(b-1)}{q}\right). \tag{2.57}$$

The nontrivial fixed point P_2 belongs to Q if its second component is nonnegative or, equivalently,

$$q \geq \frac{a(1-b)}{a-1}. \tag{2.58}$$

Provided that (2.58) holds, it is not difficult to show that the point P_2 belongs, more precisely, to the triangle

$$T = \{(x, y) : 0 \leq x \leq 1, 0 \leq y \leq a(1-x)\}.$$

It is also possible to prove, by boring but not difficult computations, that T is positively invariant for (2.56) under the following conditions:

I. $a < 4$
II. $b + q \geq a$
III. $q > b$
IV. $4q > (q+b)^2$.

The eigenvalues of the linearized system at $x = y = 0$ are a and b; hence, the origin is a saddle point. At the point P_1, the eigenvalues of the linearized system are $\lambda_1 = 2 - a$ and $\lambda_2 = b + q(a-1)/a$. Thus, $\lambda_1 < 1$ and, if $a < 3$, also $\lambda_1 > -1$. Moreover, by virtue of (2.58), $\lambda_2 \geq 1$. Hence, if the inequality in (2.58) is strict, P_1 is a saddle point, as well. Instead, if the equality holds in (2.58), then $\lambda_2 = 1$ and the system becomes critical; notice that in this case, we also have $P_1 = P_2$.

In general, the analysis of the eigenvalues placement for the nontrivial fixed point P_2 can be carried on with the aid of the so-called Schur-Cohn criterion (the discrete analogous of the Routh-Hurwitz criterion, see [3], p. 27). The Schur-Cohn criterion takes a particularly simple form when applied to second order algebraic equations with real coefficients. It states that if $|c_2| < 1$ and $|c_1| < 1 + c_2$ then the roots of the polynomial

$$\lambda^2 + c_1\lambda + c_2$$

both have modulus less than 1. In our case, if (2.58) holds and if $a < 3$, this leads to the additional condition

$$q < \frac{a(2-b)}{a-1} \tag{2.59}$$

for asymptotic stability of P_2. A rather complete and interesting investigation of the stability of P_2 can be achieved in a more direct way fixing the values of a and b and looking at q as a parameter. Let for instance $a = 3/2$ and $b = 1/2$, in which case (2.58) reduces to $q \geq 3/2$ and (2.59) to $q < 9/2$. Using again the symbols λ_1, λ_2 for the eigenvalues of the linearized system at P_2, we can summarize the conclusions in the following way.

(1) $0 < q < 15/34$. The eigenvalues are real, with $\lambda_1 < -1$ and $\lambda_2 > 1$: the system is completely unstable at P_2.
(2) $15/34 < q < 3/2$. The eigenvalues are real, with $|\lambda_1| < 1$ (λ_1 crosses the origin when $q = 9/10$) and $\lambda_2 > 1$.
(3) $3/2 < q < 3(1 + \sqrt{2})/4$. The eigenvalues are still real. Both are positive and less than 1. The point P_2 is locally asymptotically stable.
(4) $3(1 + \sqrt{2})/4 < q < 9/2$. The eigenvalues are complex conjugate, with modulus less than 1. P_2 is still locally asymptotically stable.
(5) $q > 9/2$. The modulus of the eigenvalues becomes greater than 1. The point P_2 is completely unstable.

The changes of the dynamical behavior arising when $q = 3/2$ and $q = 9/2$ are called *bifurcations*. Bifurcation theory is the object of Chap. 3.

Figure 2.11 shows two trajectories of system (2.56) with $a = 1.5$, $b = 0.5$, $q = 9/5$ attracted by P_2. For these values of a, b, q, the triangle T is positively invariant (conditions I,II,III,IV listed above are satisfied) and the eigenvalues of the linearized system at P_2 are $5/6$ an $3/4$.   ∎

Example 2.18 Under the proportional increment hypothesis, system (2.56) gives rise to the continuous time system

$$\begin{cases} \dot{x} = (a-1)x - ax^2 - xy \\ \dot{y} = (b-1)y + qxy \end{cases} \tag{2.60}$$

where, a, b, q are as in (2.56). The equilibria of (2.60) are the same as the fixed points in (2.56). Taking $a = 3/2$ and $b = 1/2$, this time we see that P_2 is locally asymptotically stable for every $q > 3/2$. Thus in particular, the dynamics of the discrete time system (2.56) and its continuous time version differ for $q > 9/2$.   ∎

Fig. 2.11 Two trajectories
of system (2.56) with
$a = 1.5, b = 0.5, q = 1.8$
attracted by the fixed point
P_2. The corresponding initial
states are
$(0.5, 0.1), (0.06, 0.4)$

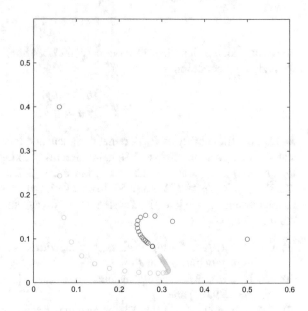

Example 2.19 This example is the discrete time version of an example presented in
[13]. It concerns a simple chemical plant formed by two cylindrical tanks in series.
The connection is realized by means of a valve placed at the bottom of the tanks. Some
amount of a fluid substance is contained in the tanks and it is supposed that some
chemical reaction is performed, while the fluid remains inside. Here, our interest is
not focused on the chemical reactions, but rather on the fluid dynamics.

A sequence of equally spaced instants of time is given. At each instant of this
sequence, the following actions are performed.

1. A volume u of the fluid is let in the first tank.
2. A volume v of the fluid is let out from the second tank, through a second valve.
3. A volume q of the fluid is let to flow from the first tank into the second one.

To model the dynamics, we introduce the following notation:

- a is the area of the horizontal section of the tanks (assuming that they have the
  same shape);
- x, y are, respectively, the hight of the column fluid in the first and second tank;
- P_1, P_2 are, respectively, the pressure in the first and second tank.

We can start writing the equations

$$\begin{cases} x_{n+1} = x_n + \dfrac{u_n - q_n}{a} \\ y_{n+1} = y_n + \dfrac{q_n - v_n}{a}. \end{cases} \tag{2.61}$$

The fluid transfers through the valves are determined by the pressure, which in turn is proportional to the hight of the column fluid; we have in particular

$$P_1 = \rho(x - y) \quad \text{and} \quad P_2 = \rho y - P_0$$

where P_0 is the atmospheric pressure and ρ is a numerical coefficient. According to Torricelli's law of fluid dynamics, we have

$$q_n = \sqrt{\rho(x_n - y_n)} \quad \text{and} \quad v_n = \sqrt{\rho y_n - P_0}. \tag{2.62}$$

Here for simplicity, we are assuming that

$$x_n \geq y_n \geq P_0/\rho \tag{2.63}$$

for each n. Moreover, we will consider the case where $u = v$, which means that the total amount of the fluid mass contained in the two tanks remains unchanged at each instant. This is equivalent to constrain the dynamics to evolve on a linear manifold of the form

$$x + y = K \tag{2.64}$$

the constant K being determined by the initial conditions. The consistency between (2.64) and (2.63) is guaranteed, provided that

$$K > \frac{2P_0}{\rho}. \tag{2.65}$$

Replacing (2.62) in (2.61) and taking into account (2.64), we finally get

$$\begin{cases} x_{n+1} = x_n + \dfrac{\sqrt{\rho y_n - P_0} - \sqrt{\rho(x_n - y_n)}}{a} \\[2mm] y_{n+1} = y_n + \dfrac{\sqrt{\rho(x_n - y_n)} - \sqrt{\rho y_n - P_0}}{a}. \end{cases} \tag{2.66}$$

System (2.66) admits infinitely many fixed points (x_*, y_*), filling the line of equation

$$y_* = \frac{1}{2}\left(x_* + \frac{P_0}{\rho}\right).$$

More precisely, for each value of $K > 2P_0/\rho$ there is a unique fixed point

$$x_* = \frac{2}{3}\left(K - \frac{P_0}{2\rho}\right), \quad y_* = \frac{1}{3}\left(K + \frac{P_0}{\rho}\right).$$

The stability of this point can be studied by the first approximation method. The Jacobian matrix of the right hand side of (2.66) at (x_*, y_*) is

$$\begin{pmatrix} 1 - \dfrac{\rho}{a\sqrt{2(\rho x_* - P_0)}} & \dfrac{\rho\sqrt{2}}{a\sqrt{\rho x_* - P_0}} \\ \dfrac{\rho}{a\sqrt{2(\rho x_* - P_0)}} & 1 - \dfrac{\rho\sqrt{2}}{a\sqrt{\rho x_* - P_0}} \end{pmatrix}.$$

The eigenvalues are easily computed: they are $\lambda_1 = 1$ (independent of K) and

$$\lambda_2 = 1 - \frac{3\rho}{a\sqrt{2(\rho x_* - P_0)}}.$$

The fixed point is therefore critical. Clearly, $\lambda_2 < 1$ for each K. In addition, we have $\lambda_2 > -1$ provided that

$$x_* > \frac{P_0}{\rho} + \frac{9\rho}{8a^2} \iff K > \frac{2P_0}{\rho} + \frac{27\rho}{16a^2}.$$

If this last condition is met, the nearby orbits lying on the manifold (2.64) are attracted by the corresponding fixed point. The convergence is monotone as far as $\lambda_2 > 0$, otherwise the system oscillates.   ∎

2.6.3   Critical Cases in Dimension One

In this section we focus on one dimensional discrete dynamical system

$$x^+ = f(x), \qquad x \in \mathbf{R} \tag{2.67}$$

under the assumption that $f \in \mathbf{C}^1(I)$, I being an open interval of $\mathbf{R}$. We also assume that there is a point x_* of I such that $f(x_*) = x_*$. According to the Theorem of stability by first approximation, if $|f'(x_*)| < 1$, x_* is locally asymptotically stable for (2.67), and if $|f'(x_*)| > 1$, x_* is completely unstable. Next proposition states in a more precise way how the local behavior of the solutions of (2.67) around x_* is determined by the linearization.

Proposition 2.10 If $0 < f'(x_*) < 1$, nearby solutions converge monotonically to the origin. If $-1 < f'(x_*) < 0$, nearby solutions converge to the origin changing sign at every new iteration (the system exhibits an oscillatory behavior).

Proof By the change of variable $y = x - x_*$, we can assume without loss of generality that $x_* = 0$. By the first order Taylor formula we can therefore write (2.67) as

$$x^+ = (f'(0) + o(1))x.$$

Hence, for each small $\varepsilon > 0$ there is a $\delta > 0$ such that if $0 < f'(0) < 1$ and $|x| < \delta$, then $\varepsilon < f'(0) + o(1) < 1 - \varepsilon$. The conclusion is straightforward.

The second statement can be proven in analogous way.   ∎

The previous proposition is illustrated for instance by the fixed point $x_* = (a - 1)/a$ of equation (2.44); if $1 < a < 2$ then $f'(x_*) > 0$ so that the solutions converge monotonically to x_* while if $2 < a < 3$ we have $f'(x_*) < 0$ and the solutions oscillate.

As suggested by Examples 2.10 and 2.11, in the critical case where $f'(x_*) = \pm 1$, the stability properties depend on the higher order terms of the Taylor expansion of $f(x)$ at x_*. Below, we present a criterion of graphic nature. For simplicity, we assume that $x_* = 0$ but the result can be easily extended to any fixed point.

Proposition 2.11 *Let the system (2.67) with $f \in \mathbf{C}^1(I)$ be given, and assume that $|f'(0)| = 1$. If there exists a number $\eta_0 > 0$ such that for every $x \in (-\eta_0, 0) \cup (0, \eta_0)$*

$$|f(x)| < |x|. \tag{2.68}$$

Then, $x_* = 0$ is locally asymptotically stable.

Proof According to (2.68) we notice that there is no fixed or periodic points (except the origin) in the interval $(-\eta_0, \eta_0)$.

Consider first the case $f'(x) = 1$. There must exist a number η_1 with $0 < \eta_1 \leq \eta_0$ such that

$$f(x) < 0 \text{ for } x \in (-\eta_1, 0) \text{ and } f(x) > 0 \text{ for } x \in (0, \eta_1).$$

Combined with (2.68) this leads to

$$x < f(x) < 0 \text{ for } x \in (-\eta_1, 0) \text{ and } 0 < f(x) < x \text{ for } x \in (0, \eta_1).$$

Now, it is not difficult to check that if $\hat{x} \in (-\eta_1, 0)$, the corresponding solution x_n is increasing and negative (hence, upper bounded). Analogously, if $\hat{x} \in (0, \eta_1)$, the corresponding solution x_n is decreasing and positive (hence, bounded from below). In both cases, the limit exists and, by continuity,

$$\ell = \lim_{n \to \infty} x_{n+1} = \lim_{n \to \infty} f(x_n) = f(\lim_{n \to \infty} x_n) = f(\ell).$$

This means that ℓ must be a fixed point of the system. It is also clear that $\ell \in (-\eta_1, \eta_1)$, and since the origin is the unique fixed point in this interval, we conclude that $\ell = 0$.

Consider now the case $f'(x) = -1$. This time, for some η_1 with $0 < \eta_1 \leq \eta_0$, we have

$$f(x) > 0 \text{ for } x \in (-\eta_1, 0) \text{ and } f(x) < 0 \text{ for } x \in (0, \eta_1)$$

which combined with (2.68) leads to

$$0 < f(x) < -x \text{ for } x \in (-\eta_1, 0) \text{ and } -x < f(x) < 0 \text{ for } x \in (0, \eta_1).$$

It is not difficult to see that for each $\hat{x} \in (-\eta_1, \eta_1)$ the sequence $|x_n|$ is decreasing, so that the limit $\lim_{n \to +\infty} |x_n| = \ell$ exists and belongs to $(-\eta_1, \eta_1)$. On the other hand, x_n changes sign at every iteration. If $\ell \neq 0$, we would have a nontrivial periodic solution of period 2. But we already noticed that this is impossible. In conclusion, $\lim_{n \to +\infty} x_n = 0$.

We have so proven that the origin is attractive in both cases. The stability property can be easily proven as a consequence of the monotonicity of $|x_n|$. ∎

Proposition 2.11 can be applied, for instance, to the system of Example 2.11. In fact, exploiting more or less the same arguments, we can also prove a criterion which does not require the differentiability assumption about f.

Proposition 2.12 *Let the system (2.67) be given, where f is continuous. Assume that there exists a number $\eta_0 > 0$ such that one of the following situations occurs.*

1. $0 \leq f(x) < |x|$ *for each $x \in (\eta_0, \eta_0)$, $x \neq 0$.*
2. $x < f(x) \leq 0$ *for each $x \in (\eta_0, 0)$ and $0 \leq f(x) < x$ for each $x \in (0, \eta_0)$.*
3. $-|x| < f(x) \leq 0$ *for each $x \in (\eta_0, \eta_0)$, $x \neq 0$.*
4. $0 \leq f(x) < -x$ *for each $x \in (\eta_0, 0)$ and $-x < f(x) \leq 0$ for each $x \in (0, \eta_0)$.*

Then, $x_* = 0$ is locally asymptotically stable.

We may also give a result about complete instability.

Proposition 2.13 *Let the system (2.67) be given, where f is continuous. Assume that there exists a number $\eta_0 > 0$ such that one of the following situations occurs.*

1. $f(x) > |x|$ *for each $x \in (\eta_0, \eta_0)$, $x \neq 0$.*
2. $f(x) < x$ *for each $x \in (\eta_0, 0)$ and $f(x) > x$ for each $x \in (0, \eta_0)$.*
3. $f(x) < -|x|$ *for each $x \in (\eta_0, \eta_0)$, $x \neq 0$.*
4. $f(x) > -x$ *for each $x \in (\eta_0, 0)$ and $f(x) < -x$ for each $x \in (0, \eta_0)$.*

Then, $x_* = 0$ is completely unstable.

Example 2.20 Consider in **R** the discrete dynamical system

$$x^+ = f(x) = \begin{cases} -\beta x & \text{if } x \le 0 \\ \\ -\alpha x & \text{if } x > 0. \end{cases}$$

Here, $0 < \alpha < 1 < \beta$. The origin is the unique fixed point. The solutions can be easily computed. If $\hat{x} < 0$, we have

$$\begin{cases} x_{2n} = \alpha^n \beta^n \hat{x} \\ x_{2n+1} = -\alpha^n \beta^{n+1} \hat{x} \end{cases}$$

and if $\hat{x} > 0$, we have

$$\begin{cases} x_{2n} = \alpha^n \beta^n \hat{x} \\ x_{2n+1} = -\alpha^{n+1} \beta^n \hat{x} \end{cases}$$

for $n = 0, 1, 2, \ldots$. Every solution changes sign at each iteration but, if $\alpha\beta < 1$, ultimately converges to the origin. The asymptotic stability of the origin can be also ascertained for instance by the following piecewise linear Lyapunov function

$$V(x) = \begin{cases} -cx & if x \le 0 \\ x & if x > 0 \end{cases}, \quad \text{with } 0 < \alpha < 1 < \beta < c < \frac{1}{\alpha}.$$

For instance, the choice $\alpha = 1/4$, $\beta = 2$, $c = 3$ works. ∎

Note that in the previous example Proposition 2.12 cannot be applied.

References

1. Biglio A.: Stabilità e Stabilizzabilità di Sistemi Non Lineari Discreti. PhD Thesis, Università di Genova, Università e Politecnico di Torino (1999) (in Italian)
2. Bacciotti, A.: Analisi della stabilità. Pitagora Editrice, Bologna (2006).(in Italian)
3. LaSalle, J.: The Stability and Control of Discrete Process. Springer Verlag, New York (1986)
4. Bhatia N.P., Szegö G.P.: Dynamical Systems: Stability Theory and Applications, Lecture Notes in Mathematics, Springer-Verlag, New York (1967)
5. Halanay, A.: Quelques questions de la thèorie de la stabilité pour les systèmes aux différences finies. Arch. Ration. Mech. Anal. **12**, 150–154 (1963)
6. Bacciotti, A., Rosier, L.: Lyapunov Functions and Stability in Control Theory, 2nd edn. Springer, Berlin (2005)
7. Bhatia, N.P., Szegö, G.P.: Stability Theory of Dynamical Systems. Springer-Verlag, Berlin (1970)
8. Hahn, W.: Stability of Motion. Springer-Verlag, Berlin (1967)
9. Hahn, W.: Theory and Applications of Lyapunov's Direct Method. Prentice-Hall, Englewoog Cliffs (1963)

10. Lang, S.: Linear Algebra. Addison Wesley, Reading Mass (1966)
11. Massera J.L.: Thèorie de la stabilité, C.I.M.E. 3° ciclo, Varenna (1954)
12. Goel, N.S., Maitra, S.C., Montroll, E.W.: Nonlinear Models of Interacting Populations. Academic Press, New York (1971)
13. Marshall, S.A.: Introduction to Control Theory. Macmillan Press, London (1978)

Chapter 3
Bifurcations

In the previous chapter we encountered several examples of systems with fixed points whose stability properties depend on the values of a parameter. For instance, for the Verhulst equation rewritten in rescaled form and state space extended to $\mathbf{R}$ (Examples 1.11, 1.14 and 2.8), the origin is a locally asymptotically stable fixed point if the parameter a belongs to $(-1, 1)$, and becomes unstable if $|a| > 1$. As already mentioned, in similar cases one is used to say that the system undergoes a *bifurcation*. When the system is defined by a function f of class at least C^1 in a neighborhood of the fixed point x_*, possible bifurcation phenomena related to displacements of the eigenvalues of the matrix $A = (Df)(x_*)$ can be classified and characterized. This is precisely the purpose of this chapter.

3.1 Diffeomorphisms

Let us recall first some useful notions. Let Ω', Ω be regions (i.e., open and connected subsets) of $\mathbf{R}^d$, and let $p \in \Omega'$, $q \in \Omega$ be given. A map $H : \Omega' \to \Omega$ is said to be a *local homeomorphism* relative to p and q if there exist a neighborhood $\mathcal{V}$ of p and a neighborhood $\mathcal{U}$ of q such that:

- $H(p) = q$;
- H is a bijection (i.e., onto and injective, and hence invertible) between $\mathcal{V}$ and $\mathcal{U}$;
- H is continuous together with its inverse map.

© The Author(s), under exclusive license to Springer Nature Switzerland AG 2022
A. Bacciotti, *Discrete Dynamics*, Mathematical Engineering,
https://doi.org/10.1007/978-3-030-95092-7_3

If $p = q$, we simply say that H is a local homeomorphism at p. If in the previous definition we can take $\mathcal{V} = \Omega'$ and $\mathcal{U} = \Omega$ we say that H is a *regional homeomorphis*, and if we have $\Omega' = \Omega = \mathcal{V} = \mathcal{U} = \mathbf{R}^d$, we say that H is a *global homeomorphism*.[1]

If in addition H is of class C^r together with its inverse map ($r \geq 1$), we say that H is a (respectively local, regional, global) C^r-*diffeomorphism*. Recall that if a map H is of class C^1 and invertible, its inverse map is not automatically of class C^1. Recall also the following facts.

Proposition 3.1 *Let $H : \Omega' \to \Omega$ be a map of class C^1 in a neighborhood of a point p, and let $H(p) = q$. Then H is a local C^1-diffeomorphism relative to p and q if and only if* $\det (DH)(p) \neq 0$.

Proposition 3.2 (Hadamard's Theorem) *Let $H : \mathbf{R}^d \to \mathbf{R}^d$ be a map of class C^1. Then, H is a global C^1-diffeomorphism if and only if the following conditions hold:*

(i) $\det (DH)(\xi) \neq 0$ *for each* $\xi \in \mathbf{R}^d$;
(ii) $\lim_{|\xi| \to \infty} |H(\xi)| = \infty$.

3.2   Stable Manifold

In Chap. 1 Sect. 1.1.3 we have introduced systems of the form

$$x^+ = f(x) , \quad x \in \Omega \subseteq \mathbf{R}^d, \quad f(\Omega) = \Omega \tag{3.1}$$

for which f is invertible on Ω. If in addition f is of class C^1 and $\det (Df)(x) \neq 0$ for each $x \in \Omega$, the analysis of the dynamical behavior can be further carried on. We limit ourselves to report some results about stable and unstable manifolds.

Consider preliminarily a linear discrete system

$$x^+ = Ax \tag{3.2}$$

where A is a real $d \times d$ square matrix. Assume that there are no eigenvalues λ of A for which $\lambda = 0$ or $|\lambda| = 1$ (i.e., A is nonsingular and the origin is a hyperbolic fixed point for (3.2)).

The following facts are well known. Let V^s be the subspace of $\mathbf{R}^d$ formed by all the generalized eigenvectors of A corresponding to some eigenvalue λ with $|\lambda| < 1$. Let moreover V^u be the subspace of $\mathbf{R}^d$ formed by all the generalized eigenvectors of A corresponding to some eigenvalue λ with $|\lambda| > 1$. V^s and V^u are called, respectively, the *stable* and *unstable* subspace of the matrix A.

---

[1] Of course, if H is a regional or global homeomorphism then it is also a local homeomorphism relative to any pair p, q, provided that $q = H(p)$.

Proposition 3.3 V^s and V^u are invariant for the linear system (3.2). Moreover,

$$V^s = \{\hat{x} \in \mathbf{R}^d : \lim_{n \to \infty} A^n \hat{x} = 0\},$$

$$V^u = \{\hat{x} \in \mathbf{R}^d : \lim_{n \to \infty} A^{-n} \hat{x} = 0\}$$

Let finally n_s be the number of the eigenvalues λ of A with $|\lambda| < 1$ (counting multiplicity), and n_u be the number of the eigenvalues λ of A with $|\lambda| > 1$ (counting multiplicity). We have $\dim V^s = n_s$, $\dim V^u = n_u$, $n_s + n_u = d$ and $V^s \oplus V^u$.

Now we come back to system (3.1), with f invertible and of class C^1 on Ω. Let x_* be a fixed point for (3.1), and let $A = (Df)(x_*)$. Assume that A is nonsingular, so that f is a local diffeomorphism at x_*. Assume also that x_* is hyperbolic i.e., there exists no eigenvalue λ of the matrix $A = (Df)(x_*)$ such that $|\lambda| = 1$ (Definition 2.15). Recall the definition of the attraction set

$$\mathcal{A}(x_*) = \{\hat{x} \in \Omega : \lim_{n \to +\infty} f^{[n]}(\hat{x}) = x_*\}$$

and define reciprocally the *repulsion set*

$$\mathcal{Z}(x_*) = \{\hat{x} \in \Omega : \lim_{n \to +\infty} f^{[-n]}(\hat{x}) = x_*\}.$$

Of course, $x_* \in \mathcal{A}(x_*) \cap \mathcal{Z}(x_*)$. $\mathcal{A}(x_*)$ and $\mathcal{Z}(x_*)$ are, respectively, positively and negatively invariant. In fact, since f is invertible, they are actually invariant.

The subspaces V^s and V^u of system (3.2) with $A = (Df)(x_*)$ are precisely the attraction set and the repulsion set relative to the origin of the linearization of (3.1). In general, V^s and V^u are not invariant for the solutions of the "exact" system (3.1). Nevertheless, these subspaces can provide useful information.

The following notation will be used: if M is a differentiable manifold in $\mathbf{R}^d$ and $x \in M$, $\mathbf{T}_x(M)$ is the tangent space of M at x. Recall that $\dim M = \dim \mathbf{T}_x(M)$ at each point $x \in M$.

Theorem 3.1 (Stable manifold) Let $f : \Omega \to \Omega$ be invertible and of class C^1. Let $x_* \in \Omega$ be a hyperbolic fixed point for (3.1) and let, in addition, $\det (Df)(x_*) \neq 0$, so that f is a local diffeomorphism at x_*.

Then, there exist a neighborhood $\mathcal{W}$ of x_* and a differentiable manifold $M^s_\mathcal{W}$ in $\mathbf{R}^d$ of class C^1 such that:

(i) $x_* \in M^s_\mathcal{W}$;
(ii) $M^s_\mathcal{W} \subseteq \mathcal{A}(x_*) \cap \mathcal{W}$;
(iii) $\dim M^s_\mathcal{W} = n^s$;
(iv) $\mathbf{T}_{x_*}(M^s_\mathcal{W}) = V^s$;
(v) $M^s_\mathcal{W}$ is positively invariant.

For the proof, the reader is referred to [1–3]. A similar theorem holds for the unstable manifold $M^u_\mathcal{W}$, with the obvious modifications (in particular, $M^u_\mathcal{W}$ turns out to be negatively invariant). We emphasize that the inclusion (ii) can be actually strict: this happens for instance in presence of a homoclinic trajectory.

The previous theorem can be extended to systems with critical fixed points. This leads to the definition of the so called center manifold or, depending on the approach, center-stable and center-unstable manifolds (see [4]).

Global versions are also available in the literature. However, the topological structure of the sets $\mathcal{A}(x_*)$ and $\mathcal{Z}(x_*)$ and of the global stable and unstable manifolds contained in these sets can be very complicated [2, 5]. For instance, if there is a homoclinic orbit then there are points $\xi \neq x_*$ such that $\xi \in \mathcal{A}(x_*) \cap \mathcal{Z}(x_*)$.

Example 3.1 Consider the system in Example 1.3. Since the system is linear, we have $\mathcal{A}(0) = V^s$ and $\mathcal{Z}(0) = V^u$. Both spaces are one dimensional: V^s is the eigenspace relative to the eigenvalue λ_1 whose modulus is less than 1, V^u is the eigenspace relative to the eigenvalue λ_2 whose modulus is greater than 1. We emphasize that V^s and V^u are formed by infinitely many orbits. On V^s, since $\lambda_1 < 0$, the orbits oscillate around the origin.   ∎

Example 3.2 Consider the system in $\mathbf{R}^2$

$$\begin{cases} x^+ = \dfrac{1}{2}x \\[2ex] y^+ = 2y - \dfrac{7}{4}x^2 \end{cases} \tag{3.3}$$

whose unique fixed point is the origin. It is easy to check that

$$y = x^2 \iff y^+ = (x^+)^2.$$

In other words, the parabola of equation $y = x^2$ is invariant for (3.3). Moreover, the component x_n of each solution goes to zero for $n \to \infty$, and setting $y = x^2$ in the second equation, we obtain $y^+ = x^2/4$. Hence if $x_n \to 0$ and (x_n, y_n) moves along the parabola, $y_n \to 0$ as well (Fig. 3.1).

The parabola of equation $y = x^2$ therefore represents the stable manifold of the system at the origin. The unstable manifold coincides with the y-axis.   ∎

3.3   Equivalence: Regional and Global Notions

Other examples of diffeomorphisms are certain changes of variables introduced in Chap. 1 (see Sect. 1.1.10). There, we already noticed that homeomorphisms and diffeomorphisms define equivalence relations on the set of systems. Here, we give more precise definitions about these notions.

Fig. 3.1 Stable manifold of equation (3.3)

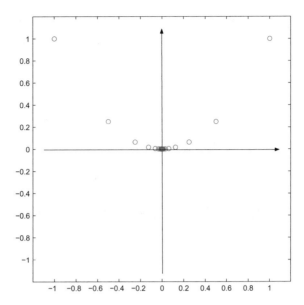

Definition 3.1 Let Ω' and Ω be as above. Let $f : \Omega \to \mathbf{R}^d$ and $g : \Omega' \to \mathbf{R}^d$ be continuous maps with $f(\Omega) \subseteq \Omega$, $g(\Omega') \subseteq \Omega'$. Consider the dynamical systems

$$x^+ = f(x) \tag{3.4}$$

and

$$y^+ = g(y). \tag{3.5}$$

We say that (3.4) and (3.5) are *regionally topologically equivalent* (or *conjugate*) if there exists a homeomorphism $x = h(y) : \Omega' \to \Omega$ such that

$$g(y) = h^{-1}(f(h(y))) \tag{3.6}$$

for all $y \in \Omega'$. We say that (3.4) and (3.5) are *globally topologically equivalent* (or *conjugate*) if in the previous definition, we can take $\Omega = \Omega' = \mathbf{R}^d$.

If in the previous definitions $h : \Omega' \to \Omega$ is a C^r-diffeomorphism ($r \geq 1$), we say that (3.4) and (3.5) are (regionally or globally) C^r-*equivalent*.

Of course, if (3.4) and (3.5) are C^r-equivalent for some $r \geq 1$, then they are also topologically equivalent. Moreover, if they are topologically equivalent and $\{y_n\}$ is the solution of (3.5) such that $y_0 = \hat{y}$, then $\{x_n = h(y_n)\}$ is the solution of (3.4) such that $x_0 = h(\hat{y})$, and vice-versa. In particular, x_* is a fixed point of (3.4) if and only if $y_* = h^{-1}(x_*)$ is a fixed point of (3.5). Analogously, x_* is a periodic point of (3.4) if and only if $y_* = h^{-1}(x_*)$ is a periodic point of (3.5).

Theorem 3.2 *Let (3.4) and (3.5), with f, g of class C^1, be regionally C^1-equivalent, and let y_* be a fixed point of (3.5). Let finally $x_* = h(y_*)$. Then, the matrices $(Df)(x_*)$ and $(Dg)(y_*)$ are similar.*

Proof For convenience, rewrite (3.6) as

$$h(g(y)) = f(h(y)).$$

Taking the derivative of both sides, we obtain

$$(Dh)(g(y))(Dg)(y) = (Df)(h(y))(Dh)(y)$$

and finally, taking into account that $g(y_*) = y_*$, we get

$$(Dg)(y_*) = (Dh)^{-1}(g(y_*))(Df)(h(y_*))(Dh)(y_*)$$
$$= (Dh)^{-1}(y_*)(Df)(x_*)(Dh)(y_*).$$

This proves the statement. ∎

Remark 3.1 In some applications, Ω and Ω' are connected and with nonempty interior, but not necessarily open. The notion of equivalence can be generalized to such situations. We will say for instance that the systems are equivalent if Definition 3.1 applies to some extensions $\tilde{f}$ and $\tilde{g}$ of f and g defined on some open sets $\tilde{\Omega}$ and $\tilde{\Omega}'$ containing respectively Ω and Ω' in their interiors. ∎

Example 3.3 Consider the linear systems in $\mathbf{R}$

$$x^+ = ax \quad \text{and} \quad y^+ = bx.$$

If $a \neq b$, Theorem 3.2 implies that these systems cannot be C^r-equivalent with $r \geq 1$. However, a topological equivalence can be established in the following cases:

1. $a, b > 1$
2. $0 < a, b < 1$
3. $-1 < a, b < 0$
4. $a, b < -1$.

For instance, in the first case the topological equivalence is realized by the transformation

$$x = h(y) = \begin{cases} y^{\log_b a} & \text{if } y \geq 0 \\ -(-y)^{\log_b a} & \text{if } y < 0. \end{cases}$$

Note that if $b < a$ then h is of class C^1, but h^{-1} is not. The same transformation works also if $0 < a, b < 1$. For the cases where a, b are negative, we may use the modified form

$$x = h(y) = \begin{cases} y^{\log_{-b}(-a)} & \text{if } y \geq 0 \\ -(-y)^{\log_{-b}(-a)} & \text{if } y < 0. \end{cases}$$

Apart from the cases listed above, the systems are not topologically equivalent. Assume for instance that $b < 0 < a$ and assume, by contradiction, that the topological equivalence holds. Let $\{y_n\}$ be a solution of the second system with, say, $y_0 > 0$. Then for some h such that $h(0) = 0$ the sequence $x_n = h(y_n)$ should be the solution of the first system such that $x_0 = h(y_0)$. But $x_n = a^n x_0$ and $y_n = b^n y_0$. Hence

$$a^n x_0 = h(b^n y_0). \tag{3.7}$$

In this equality the left-hand side does not change sign, while $y_n = b^n y_0$ is positive if n is even and negative if n is odd. Thus we see that $h(y)$ must take values of the same sign both for $y < 0$ and $y > 0$. This is impossible, since h is continuous and invertible. Analogously, if for instance $0 < a < 1 < b$, we arrive to a contradiction taking the limit for $n \to +\infty$ on both sides of (3.7). ∎

Example 3.4 Let $f(x) = 4x(1 - x) : [0, 1] \to [0, 1]$ be the rescaled logistic map considered on the interval of biological interest with parameter $a = 4$ (see (2.44)). Let moreover $g(y) = 1 - 2|y - \frac{1}{2}| : [0, 1] \to [0, 1]$ be the tent map introduced in Example 2.6. The systems defined by these maps cannot be C^1-equivalent, since the logistic map is differentiable at each point $x \in (0, 1)$, while the tent map is not differentiable for $y = \frac{1}{2}$. However, they are topologically equivalent. Let indeed

$$h(y) = \sin^2\left(\frac{\pi y}{2}\right) : [0, 1] \to [0, 1].$$

It is not difficult to check that $f \circ h = h \circ g$. It is convenient to perform the computation separately for $0 < y < \frac{1}{2}$ and $\frac{1}{2} < y < 1$. ∎

Example 3.5 The map $f(x) = 4x(1 - x) : [0, 1] \to [0, 1]$ is C^∞-equivalent to the map $g_1(y) = 1 - 2y^2 : [-1, 1] \to [-1, 1]$ (a version of the quadratic map we will consider later). The equivalence is realized by the map

$$h(y) = \frac{1 + y}{2} : [-1, 1] \to [0, 1].$$

∎

3.4 Equivalence: Local Notions

The Hartman-Grobman Theorem is a very useful result in the analysis of the local behavior of a dynamical system. It can be reviewed as an enriched form of the theorem of stability by first approximation. Here, we present a version of the Hartman-Grobman Theorem for discrete systems (see [3, 6]). To this end, we need first to "localize" the definition of systems equivalence.

Definition 3.2 Let us consider again the systems (3.4) and (3.5), and let $p \in \Omega'$, $q \in \Omega$ be given. We say that (3.4) and (3.5) are *locally topologically equivalent* relative to p and q if there exist a neighborhood $\mathcal{V}$ of p, a neighborhood $\mathcal{U}$ of q and a homeomorphism $x = h(y) : \mathcal{V} \to \mathcal{U}$ such that $q = h(p)$ and (3.6) holds for all $y \in \mathcal{V}$.

If h is a C^r-diffeomorphism, we say that (3.4) and (3.5) are *locally C^r-equivalent* relative to p and q. If $p = q$, we simply say that (3.4) and (3.5) are locally topologically equivalent or locally C^r-equivalent at p.

We stress that the previous definition is less informative with respect to Definition 3.1. For instance, in the case $d = 2$, if p and q are fixed points of the saddle type, they will be recognized as locally equivalent even if in one case but not in the other, a homoclinic orbit is present.

Theorem 3.3 (Hartman-Grobman Theorem) *Assume that x_* is a fixed point for the discrete system*

$$x^+ = f(x). \tag{3.8}$$

Assume further that the following conditions hold.

(i) $f : \Omega \to \mathbf{R}^d$ is of class at least C^1 in a neighborhood of x_*.
(ii) The matrix $(Df)(x_*)$ is nonsingular.
(iii) The fixed point x_* is hyperbolic.

Then system (3.8) is locally topologically equivalent relative to x_* and $y_* = 0$ to its linear part

$$y^+ = (Df)(x_*)y.$$

Roughly speaking, the Hartman-Grobman Theorem states that the dynamical behavior of any sufficiently regular system around a hyperbolic fixed point (and not only its stability) is completely determined by its linear part. Thus, as in the case of continuous time systems, the classification of nonlinear hyperbolic discrete dynamical systems around a fixed point reduces to the problem of classifying the behavior of linear systems around the origin. In particular, there is only finitely many topologically different situations. The following statement is given in [2].

Theorem 3.4 *Let A, B be nonsingular d × d real matrices, and assume that they do not have eigenvalues of modulus 1. The linear systems on $\mathbf{R}^d$*

$$x^+ = Ax \quad \text{and} \quad y^+ = By$$

are topologically equivalent if and only if the following conditions are met.

(i) The matrices A and B have the same number of eigenvalues inside the unit disc of the complex plane (counting multiplicities) and hence also the same number of eigenvalues outside the unit disc.

(ii) The product of the eigenvalues of A lying inside the unit disc has the same sign as the product of the eigenvalues of B lying inside the unit disc.

(iii) The product of the eigenvalues of A lying outside the unit disc has the same sign as the product of the eigenvalues of B lying outside the unit disc.

A sketch of the proof of the previous Theorem for scalar systems is suggested by the Example 3.3. Notice that since A, B are real, to evaluate the sign of the product of the eigenvalues it is sufficient to consider only the real eigenvalues.

3.5 Bifurcation Values

In this chapter the dependence on the parameters plays a crucial role. We will limit ourselves to those systems where only one scalar parameter, that we agree to denote by the symbol μ, is present. From now on, the system at hand will be therefore written as

$$x^+ = f(\mu, x) \tag{3.9}$$

where $f(\mu, x)$ is defined for each $x \in \Omega$ and each $\mu \in \mathbf{R}$. We also assume that $\Omega \subseteq \mathbf{R}^d$ is nonempty, open and connected, and that $f(\mu, \Omega) \subseteq \Omega$ for each $\mu \in \mathbf{R}$.

We are especially interested in studying how changes of the dynamics behavior in proximity of a fixed point, and in particular changes of the stability properties, can be induced by small variations of μ. The following lemma will be useful.

Lemma 3.1 *Let $x_* \in \Omega$ and $\mu_0 \in \mathbf{R}$ be given. Consider the system (3.9) with f of class C^1 with respect to both variables in a neighborhood of the point $(\mu_0, x_*) \in \mathbf{R} \times \Omega$, and assume that x_* is a fixed point when $\mu = \mu_0$, that is*

$$f(\mu_0, x_*) = x_*. \tag{3.10}$$

If 1 is not an eigenvalue of the matrix $(D_x f)(\mu_0, x_*)$, then there exists $\delta > 0$ and a unique function $x = \varphi(\mu) : (\mu_0 - \delta, \mu_0 + \delta) \to \mathbf{R}^d$ taking values in Ω such that $\varphi(\mu_0) = x_*$ and $f(\mu, \varphi(\mu)) = \varphi(\mu)$. Moreover, φ is of class C^1.

Proof The function $x = \varphi(\mu)$ we are looking for is a solution of the implicit function problem

$$F(\mu, x) = f(\mu, x) - x = 0.$$

By the Implicit Function Theorem, a function $x = \varphi(\mu)$ with the required properties exists if the matrix

$$M = (D_x F)(\mu_0, x_*) = (D_x f)(\mu_0, x_*) - I$$

is nonsingular. This is actually true. Indeed, in the opposite case, we could find a nonzero vector $v \in \mathbf{R}^d$ for which $Mv = 0$ that is

$$(D_x f)(\mu_0, x_*)v = v$$

which means that 1 is an eigenvalue of $(D_x f)(\mu_0, x_*)$. ∎

Remark 3.2 Lemma 3.1 states that if 1 is not an eigenvalue of

$$(D_x f)(\mu_0, x_*)$$

and μ is sufficiently near to μ_0, system (3.9) has one and only one fixed point in the proximity of x_*. More precisely, these fixed points are located along a differentiable curve $x = \varphi(\mu)$.

When $d = 1$, the graph of $\varphi(\mu)$ can be conveniently represented in a plane referred to the coordinates (μ, x). Moreover, as a consequence of the Implicit Function Theorem, we have

$$\varphi'(\mu_0) = \frac{(D_\mu f)(\mu_0, x_*)}{1 - (D_x f)(\mu_0, x_*)}.$$

The curve which describes the position of the fixed points looks like a single "branch" which crosses transversally the vertical x-axis.

∎

Definition 3.3 Let $x_* \in \Omega$ and $\mu_0 \in \mathbf{R}$ be given, and assume that (3.10) holds. Assume in addition that the matrix $(D_x f)(\mu_0, x_*)$ is nonsingular. We say that μ_0 is a *regular value* for (3.9) at x_* if there exists $\delta > 0$ such that for each $\mu \in (\mu_0 - \delta, \mu_0 + \delta)$ the systems

$$x^+ = f(\mu_0, x) \quad \text{and} \quad y^+ = f(\mu, y) \tag{3.11}$$

are locally topologically equivalent at x_*. We say that μ_0 is a *bifurcation value* for (3.9) at x_* if μ_0 is not a regular value.

If x_* is a hyperbolic fixed point of (3.9) for $\mu = \mu_0$, then the assumption of Lemma 3.1 is surely met and we have a curve $\varphi(\mu)$ of fixed points such that $\varphi(\mu_0) = x_*$. Moreover by continuity, taking possibly a smaller $\delta > 0$, the position of the eigenvalues of the matrix $(D_x f)(\mu, \varphi(\mu))$ does not change very much. Thus, if x_* is hyperbolic then the fixed points lying on the curve $\varphi(\mu)$ are hyperbolic, as well. Since we limit ourselves to cases where zero is not an eigenvalue of $(D_x f)(\mu_0, x_*)$, we may assume that the same holds for $(D_x f)(\mu, \varphi(\mu))$. Thus the Hartman-Grobman Theorem and Theorem 3.4 imply that μ_0 is a regular value at x_*. We can therefore state the following proposition.

Proposition 3.4 *If μ_0 is a bifurcation value at the fixed point x_*, then x_* is not hyperbolic.*

In other words, occurrence of bifurcations should be sought when x_* is critical. On the other hand, if (3.10) holds, x_* is isolated and μ_0 is regular, then for some small $\delta > 0$ and each $\mu \in (\mu_0 - \delta, \mu_0 + \delta)$ system (3.9) must have a unique fixed point.

Remark 3.3 In what follows, we will always use the term "bifurcation value" in the sense specified by Definition 3.3. However, it should be pointed out that the term "bifurcation value" is more traditionally used in the literature to identify cases where the dynamical behavior of the system for $\mu = 0$ significantly differs from that of the same system for nearby values of μ and, specifically, when several curves formed by fixed points are present for $\mu \neq \mu_0$, and at least one of the eigenvalues of the linearization of (3.9) evaluated along one of these curves crosses the boundary of the unit disc of the complex plane, determining in this way a change of stability.

As we shall see later, bifurcations are possible even if the linearization of the system at $x = x_*$, $\mu = \mu_0$ is critical, but 1 is not an eigenvalue. In these cases the bifurcation could be due, for instance, to the "birth" of periodic orbits.

∎

Remark 3.4 Definition 3.3 excludes cases where zero is an eigenvalue of $(D_x f)$ (μ_0, x_*). To comment this choice, consider the scalar linear system

$$x^+ = f(\mu, x) = \mu x$$

where we have a fixed point at the origin for each $\mu \in \mathbf{R}$. The discussion of Example 3.3 shows that, according to Definition 3.3, there are bifurcation values at $\mu = -1$, $\mu = 1$. All the other values are regular, except $\mu = 0$ which is not covered by the definition. We may notice that in any sufficiently small neighborhood of $\mu = 0$ there exist values μ_1, μ_2 such that the corresponding systems are not topologically equivalent. Nevertheless, the fixed point $x_* = 0$ remains asymptotically stable. Thus, relevant changes in the dynamics of the system when μ varies around zero are not expected.

∎

Example 3.6 For each $\mu \in \mathbf{R}$, the scalar system

$$x^+ = f(\mu, x) = \frac{x}{2} + \mu \tag{3.12}$$

has a unique fixed point $x_* = 2\mu$. One has $(D_x f)(\mu, 2\mu) = 1/2$. The previous reasoning shows that all the values of μ are regular. An explicit transformation which realizes the topological equivalence for system (3.12) for different values μ_1, μ_2 is $x = h(y) = \mu_1 y / \mu_2$. Indeed,

$$h^{-1}(f(\mu_1, h(y))) = \frac{y}{2} + \mu_2.$$

∎

The following sections provide a survey of most common types of bifurcation occurring in the case $d = 1$.

3.6   The Quadratic Map

The *quadratic map* is $f(\mu, x) = \mu - x^2 : \mathbf{R} \times \mathbf{R} \to \mathbf{R}$. It defines the dynamical system

$$x^+ = f(\mu, x) = \mu - x^2 \tag{3.13}$$

whose fixed points coincide with the real solutions of the algebraic equation

$$x^2 + x - \mu = 0.$$

Hence, we have:

- no fixed points if $\mu < -1/4$;
- a unique fixed point if $\mu = -1/4$, for $x_* = -1/2$;
- a pair of fixed points if $\mu > -1/4$, for

$$x_*^- = \frac{-1 - \sqrt{1 + 4\mu}}{2} \;, \quad x_*^+ = \frac{-1 + \sqrt{1 + 4\mu}}{2}. \tag{3.14}$$

Note that $\frac{\partial f}{\partial x}(-\frac{1}{4}, -\frac{1}{2}) = 1$, so Lemma 3.1 does not apply at $x_* = -\frac{1}{2}$, $\mu_0 = -\frac{1}{4}$. The value $\mu_0 = -1/4$ is actually a bifurcation value: indeed, when μ crosses μ_0 the number of fixed points changes, so that no topological equivalence can exist.

The curves of fixed points $x_*^-(\mu)$ and $x_*^+(\mu)$ in (3.14) are actually the branches of a parabola, whose vertex is at $(-\frac{1}{4}, -\frac{1}{2})$. This type of bifurcation, characterized by the birth of a pair of fixed points lying both on the same half-plane determined by the line $\mu = \mu_0$ is called a *saddle-node bifurcation*.

Fig. 3.2 Bifurcation diagram for the quadratic map

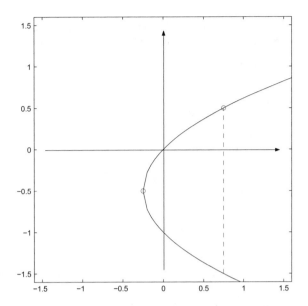

To visualize the situation, one can draw the curves $x_*^-(\mu)$ and $x_*^+(\mu)$ in the plane referred to the coordinates (μ, x). Such a representation is called the *bifurcation diagram* (see Fig. 3.2).

To complete the picture, we may investigate the stability of the bifurcated fixed points. We already noticed that for $\mu = -1/4$ the fixed point $x_* = -1/2$ is critical: in fact, it is not difficult to see that this fixed point is semi-stable. Moreover, we have that

$$\frac{\partial f}{\partial x}(\mu, x_*^+(\mu)) \in (-1, 1)$$

for $-1/4 < \mu < 3/4$. For these values of μ, the Theorem of stability by first approximation applies and we see that x_*^+ is asymptotically stable. Instead,

$$\frac{\partial f}{\partial x}(\mu, x_*^-(\mu)) > 1$$

and hence x_*^- is repulsive for each $\mu > -1/4$.

Reasonably, we may expect a new change in the system dynamics for $\mu = 3/4$. As a matter of fact, $x_*^+(3/4) = 1/2$ and $\frac{\partial f}{\partial x}(3/4, 1/2) = -1$. So, the point is not hyperbolic. However, Lemma 3.1 applies. We will continue later the discussion of this example (Example 3.9).

The saddle-node bifurcation occurs not only in the case of the quadratic map but, generically, for each one-dimensional system with the same type of criticality. In general, the curve of the bifurcated fixed points is not a parabola, but will have a similar qualitative graph.

Theorem 3.5 *Let the system (3.9) be given, with f of class C^2 with respect to both variables in a neighborhood of the point $(\mu_0, x_*) \in \mathbf{R} \times \Omega$. Assume that:*

(i) $f(\mu_0, x_*) = x_*$;

(ii) $\dfrac{\partial f}{\partial x}(\mu_0, x_*) = 1$;

(iii) $\dfrac{\partial f}{\partial \mu}(\mu_0, x_*) \neq 0$;

(iv) $\dfrac{\partial^2 f}{\partial x^2}(\mu_0, x_*) \neq 0$.

Then, there exist a number $\varepsilon > 0$ and a unique map $\mu = \psi(x) : (x_* - \varepsilon, x_* + \varepsilon) \to \mathbf{R}$ such that $\mu_0 = \psi(x_*)$ and

$$f(\psi(x), x) = x$$

for each $x \in (x_* - \varepsilon, x_* + \varepsilon)$. The map $\mu = \psi(x)$ is of class C^1 and $\psi'(x_*) = 0$. Moreover, for each $x \in (x_* - \varepsilon, x_* + \varepsilon) \setminus \{x_*\}$ either $\psi(x) < \mu_0$ or $\psi(x) > \mu_0$. Finally, the fixed points $(\psi(x), x)$ with $x \in (x_* - \varepsilon, 0)$ have opposite stability properties with respect to the fixed points $(\psi(x), x)$ with $x \in (0, x_* + \varepsilon)$.

Remark 3.5 In the previous theorem, assumption (i) states that x_* is a fixed point for (3.9) when $\mu = \mu_0$ and assumption (ii) specifies its criticality. Note that they are expressed by an equality. On the contrary, assumptions (iii) and (iv) are expressed by inequalities, and so they are generic in the sense that they are preserved under sufficiently small perturbations of the coefficients of the system and/or alterations of the higher order terms of the Taylor expansion.

■

Proof of Theorem 3.5 Let $F(\mu, x) = f(\mu, x) - x$, and consider the implicit function problem

$$F(\mu, x) = 0.$$

By (i), this equation is satisfied for $x = x_*$ and $\mu = \mu_0$. Clearly,

$$\frac{\partial F}{\partial \mu}(\mu_0, x_*) = \frac{\partial f}{\partial \mu}(\mu_0, x_*) \neq 0$$

because of (iii). Hence, the Implicit Function Theorem guarantees the existence of a number $\varepsilon > 0$ and a C^2 function $\mu = \psi(x) : (x_* - \varepsilon, x_* + \varepsilon) \to \mathbf{R}$ such that $\psi(x_*) = \mu_0$ and

$$F(\psi(x), x) = f(\psi(x), x) - x = 0 \tag{3.15}$$

for each $x \in (x_* - \varepsilon, x_* + \varepsilon)$, which means in particular that any $\bar{x} \in (x_* - \varepsilon, x_* + \varepsilon)$ is a fixed point of (3.9) when $\mu = \psi(\bar{x})$. Next we examine the qualitative shape of $\psi(x)$. From (3.15) we obtain, by derivation,

$$\frac{\partial f}{\partial x}(\psi(x), x) + \frac{\partial f}{\partial \mu}(\psi(x), x)\psi'(x) - 1 = 0 \tag{3.16}$$

for each $x \in (x_* - \varepsilon, x_* + \varepsilon)$. By (ii), when $x = x_*$ and $\mu = \mu_0$, (3.16) reduces to

$$\frac{\partial f}{\partial \mu}(\mu_0, x_*)\psi'(x_*) = 0$$

and because of (iii), we conclude that $\psi'(x_*) = 0$. Then by a further derivation, from (3.16) we obtain

$$\frac{\partial^2 f}{\partial x^2}(\psi(x), x) + \psi'(x)\left[2\frac{\partial^2 f}{\partial x \partial x}(\psi(x), x) + \frac{\partial^2 f}{\partial \mu^2}(\psi(x), x)\psi'(x)\right]$$
$$+ \frac{\partial f}{\partial \mu}(\psi(x), x)\psi''(x) = 0.$$

Since $\psi'(x_*) = 0$, when $x = x_*$ and $\mu = \mu_0$ we get

$$\frac{\partial^2 f}{\partial x^2}(\mu_0, x_*) + \frac{\partial f}{\partial \mu}(\mu_0, x_*)\psi''(x_*) = 0$$

and finally, using again (iii), we find

$$\psi''(x_*) = -\frac{\dfrac{\partial^2 f}{\partial x^2}(\mu_0, x_*)}{\dfrac{\partial f}{\partial \mu}(\mu_0, x_*)}$$

which is not zero by virtue of (iv). In conclusion, we can state that $\mu = \psi(x)$ has a local strict minimum or a local strict maximum at x_*.

To complete the proof, we have to discuss the stability properties of the bifurcated fixed points. Taking possibly a smaller $\varepsilon > 0$, (iii) yields

$$\frac{\partial f}{\partial \mu}(\psi(x), x) \neq 0 \tag{3.17}$$

for each $x \in (x_* - \varepsilon, x_* + \varepsilon)$. Thus, the left-hand side of (3.17) does not change sign. We limit ourselves to the case where $\mu = \psi(x)$ has a minimum (for the case of the maximum the proof is analogous). Thus, for a sufficiently small $\delta > 0$ and any $\bar{\mu} \in (\mu_0, \mu_0 + \delta)$ there exist two points $\xi < x_* < \zeta$ such that $\psi(\xi) = \psi(\zeta) = \bar{\mu}$. Moreover,

$$\psi'(\xi) < 0 \quad \text{while} \quad \psi'(\zeta) > 0. \tag{3.18}$$

We are now ready to apply the Theorem of stability by first approximation to the system

$$x^+ = f(\bar{\mu}, x)$$

at the fixed points ξ and ζ. To this end, we need to compute

$$\frac{\partial f}{\partial x}(\bar{\mu}, \xi) \quad \text{and} \quad \frac{\partial f}{\partial x}(\bar{\mu}, \zeta).$$

We can proceed in the following way. Let us set $x = \zeta$ in (3.16). We get

$$\frac{\partial f}{\partial x}(\bar{\mu}, \zeta) = 1 - \frac{\partial f}{\partial \mu}(\bar{\mu}, \zeta)\psi'(\zeta). \tag{3.19}$$

Instead, setting $x = \xi$, we have

$$\frac{\partial f}{\partial x}(\bar{\mu}, \xi) = 1 - \frac{\partial f}{\partial \mu}(\bar{\mu}, \xi)\psi'(\xi). \tag{3.20}$$

The comparison between (3.19) and (3.20) leads to the conclusion, taking into account of (3.18) and recalling that the left-hand side of (3.17) does not change sign.
∎

Remark 3.6 If we image that μ is moved on its axis from right to left, the saddle-node bifurcation phenomenon can be described as the collision of a pair of fixed points: one of them is characterized by an eigenvalue greater than 1 (unstable), the other by an eigenvalue less than 1 (asymptotically stable). At the instant of collision, the eigenvalue is exactly 1 (critical). After the collision, the fixed points disappear.
∎

3.7 The Transcritical Bifurcation

The standard example for the next type of bifurcation is the system

$$x^+ = f(\mu, x) = \mu x - x^2. \tag{3.21}$$

For each $\mu \in \mathbf{R}$, we have two fixed points: the origin and $x_* = \mu - 1$. When $\mu = 1$ the fixed points coincide. We have

$$\frac{\partial f}{\partial x}(1, 0) = 1,$$

so that in particular the origin is a critical point for $\mu = 1$. Note also that

$$\frac{\partial f}{\partial \mu}(1,0) = 0$$

so this case is not covered by Theorem 3.5. More generally, we have

$$\frac{\partial f}{\partial x}(\mu, 0) = \mu$$

and

$$\frac{\partial f}{\partial x}(\mu, \mu - 1) = 2 - \mu.$$

We therefore see that the origin is asymptotically stable if $-1 < \mu < 1$ while $x_* = \mu - 1$ is asymptotically stable if $1 < \mu < 3$. In particular, when μ crosses the value 1 from left to right, the stability property is lost by the origin and simultaneously acquired by x_*. Since topological equivalence preserves stability, we conclude that $\mu = 1$ is a bifurcation value at the origin.

A bifurcation represented by the intersection of two curves of fixed points with exchange of stability is called *transcritical*. For the case of Eq. (3.21), the bifurcation diagram is shown in Fig. 3.3.

Next theorem states the generic conditions under which a transcritical bifurcation occurs.

Theorem 3.6 *Let the system (3.9) be given, with f of class C^3 with respect to both variables in a neighborhood of $(\mu_0, x_*) \in \mathbf{R} \times \Omega$. Assume that:*

Fig. 3.3 Bifurcation diagram for Eq. (3.21)

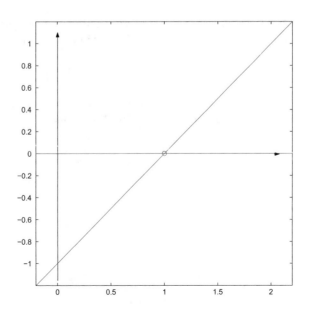

(i) $f(\mu, x_*) = x_*$ for each $\mu \in \mathbf{R}$;

(ii) $\dfrac{\partial f}{\partial x}(\mu_0, x_*) = 1$;

(iii) $\dfrac{\partial^2 f}{\partial x \partial \mu}(\mu_0, x_*) \neq 0$;

(iv) $\dfrac{\partial^2 f}{\partial x^2}(\mu_0, x_*) \neq 0$.

Then, there exist a number $\delta > 0$ and a unique map $x = \kappa(\mu) : (\mu_0 - \delta, \mu_0 + \delta) \to \mathbf{R}$ such that $x_* = \kappa(\mu_0)$ and

$$f(\mu, \kappa(\mu)) = \kappa(\mu)$$

for each $\mu \in (\mu_0 - \delta, \mu_0 + \delta)$. The map $x = \kappa(\mu)$ is of class C^1 and $\kappa'(\mu_0) \neq 0$. Moreover, the fixed point x_* has opposite stability properties for $\mu < \mu_0$ and $\mu > \mu_0$.

Remark 3.7 In the previous theorem, assumption (i) states that x_* is a fixed point for (3.9) for <u>each</u> μ. It implies in particular that $\frac{\partial f}{\partial \mu}(\mu, x_*) = 0$ for each μ. Assumption (ii) implies that Lemma 3.1 cannot be applied. Assumptions (i) and (ii) are expressed by equalities while assumptions (iii) and (iv) are generic.

∎

Proof of Theorem 3.6 The structure of the proof is similar to that of Theorem 3.5, but a preliminary application of Hadamard's Lemma is required. It allows us to represent $f(\mu, x)$ in the form

$$f(\mu, x) = f(\mu_0, x_*) + (x - x_*)g_1(\mu, x) + (\mu - \mu_0)g_2(\mu, x) \qquad (3.22)$$

where g_1, g_2 are some C^2 functions defined in some neighborhood of (μ_0, x_*). Taking into account (i), we have the additional information that

$$g_2(\mu, x_*) = 0 \qquad (3.23)$$

for each μ in neighborhood of μ_0. This in turn implies that $g_2(\mu, x) = (x - x_*)h(\mu, x)$ where h is some C^1 function. Summing up, (3.22) becomes

$$f(\mu, x) = x_* + (x - x_*)[g_1(\mu, x) + (\mu - \mu_0)h(\mu, x)]. \qquad (3.24)$$

Now we compute the partial derivatives of f. Using (i), (ii), (iv) and (3.23) we obtain:

$$\frac{\partial f}{\partial x}(\mu_0, x_*) = g_1(\mu_0, x_*) = 1; \qquad (3.25)$$

$$\frac{\partial^2 f}{\partial x^2}(\mu_0, x_*) = 2\frac{\partial g_1}{\partial x}(\mu_0, x_*) \neq 0; \qquad (3.26)$$

$$\frac{\partial^2 f}{\partial x \partial \mu}(\mu_0, x_*) = \frac{\partial g_1}{\partial \mu}(\mu_0, x_*) + h(\mu_0, x_*) \neq 0. \tag{3.27}$$

Now we are ready to examine the fixed points of the system. The equation

$$x - x_* = f(\mu, x) - x_* = (x - x_*)[g_1(\mu, x) + (\mu - \mu_0)h(\mu, x)]$$

splits into the two equations $x = x_*$ (constant solution) and

$$G(\mu, x) = g_1(\mu, x) + (\mu - \mu_0)h(\mu, x) = 1. \tag{3.28}$$

From (3.25) it follows that (3.28) is satisfied for $x = x_*$ and $\mu = \mu_0$. Moreover, from (3.26)

$$\frac{\partial G}{\partial x}(\mu_0, x_*) = \frac{\partial g_1}{\partial x}(\mu_0, x_*) \neq 0.$$

Hence, by the Implicit Function Theorem, there is a function $x = \kappa(\mu)$ defined in some neighborhood of μ_0 such that $\kappa(\mu_0) = x_*$ and

$$G(\mu, \kappa(\mu)) = 1.$$

Moreover, using (3.26) and (3.27) we can check that $\kappa'(\mu_0) \neq 0$, so that the graph of κ intersects transversally the μ-axis.

To complete the proof, we remark that because of (ii) and (iii), the value of $\frac{\partial f}{\partial x}(\mu, x_*)$ leaves/enters the interval $(-1, 1)$ when μ crosses the bifurcation value μ_0. ∎

Example 3.7 With the notation of this section, the well known rescaled logistic equation with state space $x \in \mathbf{R}$ writes

$$x^+ = f(\mu, x) = \mu x(1 - x). \tag{3.29}$$

We already know that there are two fixed points for each μ: the origin and $x_* = 1 - 1/\mu$. The assumptions of Theorem 3.6 are met with $x_* = 0$ and $\mu_0 = 1$. Thus, we have a transcritical bifurcation. As already known, the origin is asymptotically stable for $\mu < 1$ and becomes unstable for $\mu > 1$. The bifurcation diagram is shown in Fig. 3.4. ∎

Remark 3.8 There is a slightly improved form of Theorem 3.6. Indeed, Assumption (i) can be replaced by

(i') there exist $\delta_0 > 0$ and a C^1 function $x = \varphi(\mu) : (\mu_0 - \delta_0, \mu_0 + \delta_0) \to \mathbf{R}$ such that $\varphi(\mu_0) = x_*$ and $f(\mu, \varphi(\mu)) = \varphi(\mu)$ for each $\mu \in (\mu_0 - \delta_0, \mu_0 + \delta_0)$.

Fig. 3.4 Bifurcation
diagram for Eq. (3.29)

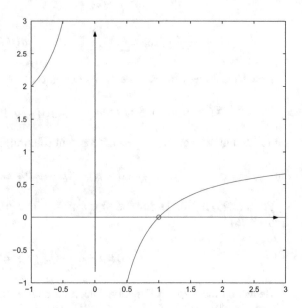

This extension is proven by the simple remark that if the map $f(\mu, x)$ satisfies
(i') in addition to (ii), (iii), (iv) of Theorem 3.6, then the map

$$F(\mu, z) = f(\mu, z + \varphi(\mu)) - \varphi(\mu)$$

satisfies (i), (ii), (iii), (iv) of Theorem 3.6 with $x_* = 0$. Finally, from the identity

$$F(\mu, \kappa(\mu)) = \kappa(\mu)$$

we deduce

$$f(\mu, \tilde{\kappa}(\mu)) = \tilde{\kappa}(\mu)$$

where $\tilde{\kappa}(\mu) = \kappa(\mu) + \varphi(\mu)$.

■

Example 3.8 The system

$$x^+ = f(\mu, x) = \mu + \mu x - x^2 - 2\mu(\mu - x) \tag{3.30}$$

has a line of fixed points $x = \mu$. In particular, for $x_* = \mu_0 = 1$ the assumptions (i'),
(ii), (iii), (iv) are satisfied (the substitution $z = x - \mu$ reduces (3.30) to (3.21)). The
second line of fixed points passing through the point $x_* = \mu_0 = 1$ is $x = 2\mu - 1$.
We may notice that none of these lines can be obtained from the fixed point equation
$f(\mu, x) - x = 0$ on the base of Lemma 3.1 since $x_* = 1$ is critical. ■

3.8 The Pitchfork Bifurcation

To introduce the next type of bifurcation, we consider the system

$$x^+ = f(\mu, x) = \mu x - x^3 = x(\mu - x^2) \tag{3.31}$$

with $x \in \mathbf{R}$. The fixed points search leads to the equation $x = x(\mu - x^2)$ whose solutions are $x = 0$ for each $\mu \in \mathbf{R}$ and $x = \pm\sqrt{\mu - 1}$ provided that $\mu \geq 1$. In other words, we have a unique fixed point for $\mu \leq 1$, and three distinct fixed points if $\mu > 1$. The fixed points with $x = \pm\sqrt{\mu - 1}$ form the branches of a parabola (see Fig. 3.5). This type of bifurcation is called a *pitchfork bifurcation*. Moreover, we have

$$\frac{\partial f}{\partial x}(\mu, 0) = \mu$$

so that the origin is asymptotically stable when $-1 < \mu < 1$ and becomes unstable when $\mu > 1$. In particular the origin is critical for $\mu = 1$, and $\mu = 1$ is a bifurcation value.

As far as the other fixed points are concerned (for $\mu > 1$), we have

$$\frac{\partial f}{\partial x}(\mu, \pm\sqrt{\mu - 1}) = 3 - 2\mu.$$

Therefore, they have the same stability properties. More precisely, they are asymptotically stable if $1 < \mu < 2$. The following theorem states the genericity conditions concerning the pitchfork bifurcation.

Theorem 3.7 *Let the system (3.9) be given, with f of class C^4 with respect to both variables in a neighborhood of $(\mu_0, x_*) \in \mathbf{R} \times \Omega$. Assume that:*

(i) $f(\mu, x_*) = x_*$ *for each* $\mu \in \mathbf{R}$;

(ii) $\dfrac{\partial f}{\partial x}(\mu_0, x_*) = 1$;

(iii) $\dfrac{\partial^2 f}{\partial x^2}(\mu_0, x_*) = 0$;

(iv) $\dfrac{\partial^2 f}{\partial x \partial \mu}(\mu_0, x_*) \neq 0$;

(v) $\dfrac{\partial^3 f}{\partial x^3}(\mu_0, x_*) \neq 0$.

Then, there exist a number $\varepsilon > 0$ and a unique map $\mu = \psi(x) : (x_* - \varepsilon, x_* + \varepsilon) \to \mathbf{R}$ such that $\psi(x_*) = \mu_0$ and

$$f(\psi(x), x) = x$$

Fig. 3.5 Bifurcation
diagram for equation (3.31)

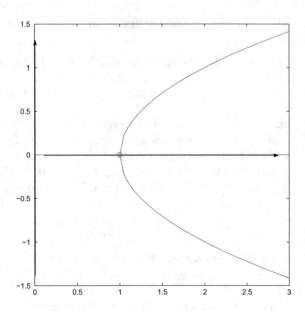

for each $x \in (x_* - \varepsilon, x_* + \varepsilon)$. The map $\mu = \psi(x)$ is of class C^1 and $\psi'(x_*) = 0$. Moreover, for each $x \in (x_* - \varepsilon, x_* + \varepsilon) \setminus \{x_*\}$ either $\psi(x) < \mu_0$ or $\psi(x) > \mu_0$. Finally, the fixed point x_* has opposite stability properties for $\mu < \mu_0$ and $\mu > \mu_0$.

Sketch of the proof As in the proof of Theorem 3.6, we begin by writing

$$f(\mu, x) = x_* + (x - x_*)[g_1(\mu, x) + (\mu - \mu_0)h(\mu, x)]$$

where now $g_1 \in C^3$ and $h \in C^2$. Then we address the problem of solving the equation

$$G(\mu, x) = g_1(\mu, x) + (\mu - \mu_0)h(\mu, x) = 1. \qquad (3.32)$$

At this point however, we cannot continue on the same way as in the proof of Theorem 3.6, since by (iii)

$$\frac{\partial G}{\partial x}(\mu_0, x_*) = \frac{\partial g_1}{\partial x}(\mu_0, x_*) = \frac{1}{2}\frac{\partial^2 f}{\partial x^2}(\mu_0, x_*) = 0.$$

However,

$$\frac{\partial G}{\partial \mu}(\mu_0, x_*) = \frac{\partial g_1}{\partial \mu}(\mu_0, x_*) + h(\mu_0, x_*) = \frac{\partial^2 f}{\partial x \partial \mu}(\mu_0, x_*) \neq 0.$$

Hence, we can recover from (3.32) a function $\mu = \psi(x)$ of class C^2 in a neighborhood of x_* such that $G(\psi(x), x) = 1$ and $\mu_0 = \psi(x_*)$. We also have $\psi'(x_*) = 0$ and, by virtue of (v), $\psi''(x_*) \neq 0$. ∎

Remark 3.9 As in the case of Theorem 3.6 and with the same argument, condition
(i) of Theorem 3.7 can be replaced by (i') (see Remark 3.8).   ∎

3.9 The Flip Bifurcation

The bifurcations presented so far have a common feature: the appearance/
disappearance of fixed points accompanied by exchange of stability occurs when
the eigenvalue of the linearized system crosses the right endpoint stability interval
$(-1, 1)$. These types of bifurcation have their analogues in the continuous time case.
This section deals with a different type of bifurcation. It originates by the entry/exit
of the eigenvalue of the linearized system through the left endpoint of the interval
$(-1, 1)$. It does not look like the previous bifurcations, since now Lemma 3.1 does
apply, and we do not have new branches of fixed points. Rather, a pair of periodic
points appears. Notice that when the eigenvalue approaches -1 from the right, the
fixed point is attractive but the solutions oscillate: this indicates that the system is
ready to bifurcate. The name of this bifurcation is *flip* or *period doubling* bifurcation.

Theorem 3.8 *Let the system (3.9) be given, with f of class C^4 with respect to both
variables in a neighborhood of $(\mu_0, x_*) \in \mathbf{R} \times \Omega$. Assume that:*

(i) $f(\mu_0, x_*) = x_*$;

(ii) $\dfrac{\partial f}{\partial x}(\mu_0, x_*) = -1$;

(iii) $\dfrac{\partial f}{\partial \mu}(\mu_0, x_*)\dfrac{\partial^2 f}{\partial x^2}(\mu_0, x_*) + 2\dfrac{\partial^2 f}{\partial x \partial \mu}(\mu_0, x_*) \neq 0$;

(iv) $3\left(\dfrac{\partial^2 f}{\partial x^2}(\mu_0, x_*)\right)^2 + 2\dfrac{\partial^3 f}{\partial x^3}(\mu_0, x_*) \neq 0$.

Then, the following statements hold.

(a) There exist a number $\delta > 0$ and a unique C^1 function $x = \varphi(\mu) : (\mu_0 - \delta, \mu_0 + \delta) \to \mathbf{R}$ such that $\varphi(\mu_0) = x_*$ and $f(\mu, \varphi(\mu)) = \varphi(\mu)$.

(b) There exist a number $\varepsilon > 0$ and a unique C^1 function $\mu = \psi(x) : (x_* - \varepsilon, x_* + \varepsilon) \to \mathbf{R}$ such that $\psi'(x_*) = 0$ and for each $x \in (x_* - \varepsilon, x_* + \varepsilon) \setminus \{x_*\}$ either $\psi(x) < \mu_0$ or $\psi(x) > \mu_0$. The points ξ and ζ such that $\psi(\xi) = \psi(\zeta)$ are periodic of period 2 for system (3.9).

*The stability properties of the fixed point $(\mu, \varphi(\mu))$ change when μ crosses the
value μ_0. On the side where the periodic points exist, they have opposite stability
properties with respect to the fixed point.*

Proof Statement (a) is nothing else that Lemma 3.1. Statement (b) results from the
application of the pitchfork bifurcation theorem to the second iterate $f^{[2]}$. Indeed,
simple calculations show that if f satisfies the conditions (i), (ii), (iii), (iv), then $f^{[2]}$
fulfills the conditions (i'), (ii), (iii), (iv), (v) of Theorem 3.7.   ∎

Fig. 3.6 Bifurcation
diagram for equation (3.33)
at $\mu_0 = 3/4$, $x_* = 1/2$
(dotted line)

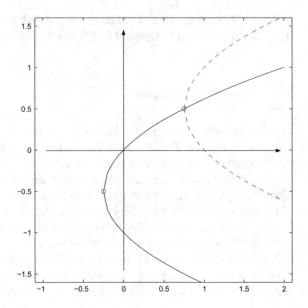

Example 3.9 We already pointed out that the system defined by the quadratic map

$$x^+ = f(\mu, x) = \mu - x^2 \tag{3.33}$$

has a branch $x_*^+(\mu) = \frac{-1 + \sqrt{1+4\mu}}{2}$ of fixed points if $\mu > -1/4$. In particular, for $\mu_0 = 3/4$ the fixed point is $x_* = 1/2$ and, for these values of μ_0 and x_*, $\frac{\partial f}{\partial x}(\mu_0, x_*) = -1$. It is not difficult to check that also the other conditions of Theorem 3.8 are fulfilled. Hence, the system has a flip bifurcation. The branches of bifurcated periodic points for (3.33) are displayed in Fig. 3.6.

∎

Other examples of flip bifurcations are provided by:

- The system (3.21) both for $\mu = -1$, $x_* = 0$ and $\mu = 3$, $x_* = 2$;
- The logistic equation (3.29) with $\mu = 3$, $x_* = 2/3$;
- The system (3.31) with $\mu = 2$, $x_* = 1$.

3.10   The Neimark-Sacker Bifurcation

Our review of local bifurcations for discrete dynamical systems ends with a description of the so-called *Neimark-Sacker bifurcation*. It involves two-dimensional systems whose linearization has a pair of complex (non real) conjugate eigenvalues. In some sense, it can be interpreted as the analog of the Hopf bifurcation in the continuous time case (in fact, it is sometimes also called *secondary Hopf bifurcation*).

Example 3.10 Recall that the delayed logistic equation (Example 2.16) can be written as a two-dimensional system

$$\begin{cases} x_{n+1} = \mu x_n (1 - y_n) \\ y_{n+1} = x_n \end{cases} \tag{3.34}$$

where, in conformity with the notation of this chapter, the coefficient of the first equation has been re-denoted by μ. Let us focus on the fixed point

$$P_\mu = \left(1 - \frac{1}{\mu}, 1 - \frac{1}{\mu} \right)$$

and recall the following information already obtained on the base of the Theorem of stability by first approximation.

- If $\frac{5}{4} < \mu < 2$ the linearization of the system at P_μ has a pair of complex conjugate eigenvalues

$$\lambda_{1,2} = \frac{1 \pm i \sqrt{4\mu - 5}}{2}$$

lying in the interior of the unit disc, so that P_μ is asymptotically stable.
- If $\mu > 2$ the modulus of $\lambda_{1,2}$ becomes greater than 1 and P_μ becomes unstable

Note that the real part of P_μ does not depends on μ. In other words, when μ crosses the value $\mu_0 = 2$ from left to right, $\lambda_{1,2}$ move away from the real axis in opposite directions and exit from the unit disc (see Fig. 3.7). Simulations indicate that for $\mu = 2$, the fixed point is critical, but still asymptotically stable (due to the higher order terms).

Clearly, due to the change of stability properties, if $\mu_1 < 2 < \mu_2$ the systems obtained from (3.34) for $\mu = \mu_1$ and $\mu = \mu_2$, are not topologically equivalent. Hence, $\mu_0 = 2$ is a bifurcation value.

Since 1 is not an eigenvalue of the linearized system for $\mu = 2$, Lemma 3.1 applies: the bifurcation does not consist in the appearance of new fixed points. Rather, the bifurcated solutions lie on a closed curve, surrounding the fixed point. This curve is an attractive set, at least for $\mu - 2$ small (see Fig. 3.8). The diameter of the closed curve increases, when μ increases (see Fig. 3.9). ∎

Example 3.11 A similar situation can be recognized studying the system

$$\begin{cases} x_{n+1} = \frac{3}{2} x_n (1 - x_n) - x_n y_n \\ \\ y_{n+1} = \frac{1}{2} y_n + \mu x_n y_n \end{cases} \tag{3.35}$$

Fig. 3.7 Boundary crossing
of the eigenvalues of the
linearized equation (3.34) at
$\mu_0 = 2$

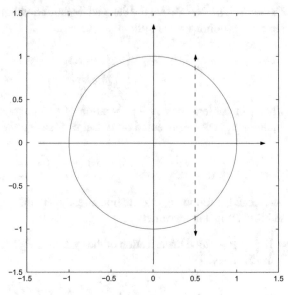

Fig. 3.8 Closed curve
covered by solutions of the
system (3.34) with
$\mu_0 = 2.05$

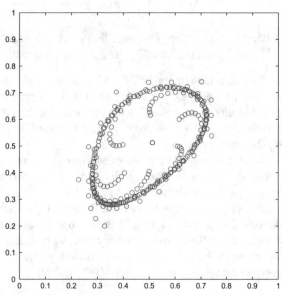

Fig. 3.9 Closed curves covered by solutions of the system (3.34) for $\mu = 2.05$ and $\mu = 2.1$

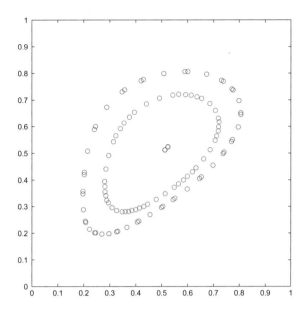

(compare with (2.56)) around the fixed point

$$\left(\frac{1}{2\mu}, \frac{1}{2} - \frac{3}{4\mu}\right)$$

which is critical for $\mu_0 = 9/2$ (see Fig. 3.10).   ∎

We now try to formalize the situations illustrated by the previous examples. Let us consider first the discrete system in polar coordinates

$$\begin{cases} \rho^+ = (1 + \mu)\rho + \alpha(\mu)\rho^3 \\ \theta^+ = \theta + \gamma(\mu) \end{cases} \tag{3.36}$$

where the functions $\alpha(\mu)$, $\gamma(\mu)$ are continuous in a neighborhood of the value $\mu_0 = 0$. We notice that the system is formed by a pair of decoupled equations. Therefore, they can be investigated separately.

First equation. Let us assume that $\alpha(0) \neq 0$. In fact, we may assume $\alpha(0) < 0$ (which implies $\alpha(\mu) < 0$ for small μ) since the opposite case is analogous. The first equation has a fixed point for $\rho = 0$. It is clear that this point is asymptotically stable if $\mu < 0$, and repulsive if $\mu > 0$. For $\mu = 0$, because of the assumption $\alpha(0) < 0$, the origin is still asymptotically stable (compare with Example 2.11).

The change of stability at the origin in the radial direction suggests that the system (3.36) undergoes a bifurcation when μ crosses the value $\mu_0 = 0$.

It is worth noticing that the first equation of (3.36) has a second fixed point

Fig. 3.10 Closed curve of bifurcated solutions of the system (3.35) for $\mu = 4.6$

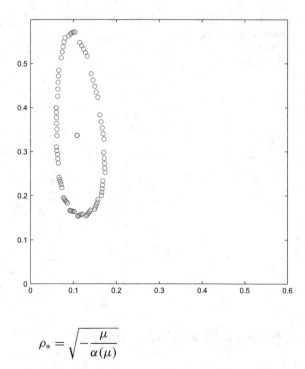

$$\rho_* = \sqrt{-\frac{\mu}{\alpha(\mu)}}$$

for $\mu > 0$. This means that the circumference centered at the origin and of radius ρ_* is invariant for the system (3.36). By linearization, we see that ρ_* is asymptotically stable when $\mu > 0$ (that is, when the origin is repulsive).

Second equation. This equation represents the well known rotation of an angle $\gamma(\mu)$ (Example 1.16): to avoid the trivial case, we will assume $\gamma(0) \neq 0$ (say $\gamma(0) > 0$ to have counterclockwise rotation). We may therefore distinguish the following situations:

• if $\gamma(\mu)/2\pi$ is rational the invariant circumference contains infinitely many periodic orbits;

• otherwise, the invariant circumference contains infinitely many dense orbits.

In both cases, the circumference is formed by the union of bifurcated orbits. If γ is not constant, the two situations above alternate in a dense way.

Now, we try to relate, through several steps, the bifurcation presented by the system (3.36) to those described in the previous examples. Consider the two-dimensional system in cartesian coordinates

$$\begin{pmatrix} x^+ \\ y^+ \end{pmatrix} = \begin{pmatrix} \cos\gamma(\mu) & -\sin\gamma(\mu) \\ \sin\gamma(\mu) & \cos\gamma(\mu) \end{pmatrix} \left[(1+\mu) \begin{pmatrix} x \\ y \end{pmatrix} \right.$$

$$\left. + (x^2 + y^2) \begin{pmatrix} \alpha(\mu) & -\beta(\mu) \\ \beta(\mu) & \alpha(\mu) \end{pmatrix} \begin{pmatrix} x \\ y \end{pmatrix} \right] \tag{3.37}$$

where, α, γ are as before and β is any continuous function. We aim to rewrite (3.37) in polar coordinates, setting

$$\rho = \sqrt{x^2 + y^2}\,, \quad \theta = \operatorname{arctg} \frac{y}{x}$$

which, apart for some obvious restrictions on θ, is equivalent to set $x = \rho\cos\theta$, $y = \rho\sin\theta$. The task is more complicated than the analogous procedure for the case of differential systems and requires in general the use of Taylor expansions. After some computations, we find

$$\rho^+ = (1+\mu)\rho\sqrt{1 + \frac{2\alpha(\mu)}{1+\mu}\rho^2 + \Phi(\mu)\rho^4}$$

$$= (1+\mu)\rho\left[1 + \frac{\alpha(\mu)}{1+\mu}\rho^2 + \frac{1}{2}\left(\Phi(\mu) - \frac{\alpha^2(\mu)}{(1+\mu)^2}\right)\rho^4 + \cdots\right]$$

where $\Phi(\mu)$ is a new function. Eliminating the terms containing powers of ρ of order strictly higher than 3, we get the first equation of (3.36). We also find

$$\theta^+ = \operatorname{arctg} \frac{y^+}{x^+} = \operatorname{arctg} \frac{(1+\mu)\sin(\theta + \gamma(\mu)) + \rho^2(\ldots)}{(1+\mu)\cos(\theta + \gamma(\mu)) + \rho^2(\ldots)}.$$

Eliminating the terms containing powers of ρ of order 2 and more, we get

$$\theta^+ = \theta + \gamma(\mu)$$

which is the second equation of (3.36). Note that in the first equation is necessary to retain the third order term since we are interested in a critical case, not decidable by the first approximation.

We continue our way, going back to the system (3.34). First of all, we replace the parameter by $\nu = \mu - 2$ and operate the change of coordinates

$$u = x - \frac{\nu+1}{\nu+2}\,, \quad v = y - \frac{\nu+1}{\nu+2}.$$

System (3.34) takes the new form

$$\begin{cases} u^+ = u - (\nu+1)\nu - (\nu+2)u\nu \\ v^+ = u. \end{cases} \tag{3.38}$$

The fixed point of (3.34) which undergoes the bifurcation has been moved to the origin (for each value of the parameter) and it becomes critical when $v = 0$. We are now in a more convenient position to compare our system and (3.37). The jacobian matrix of (3.38) at $u = v = 0$ is[2]

$$A = \begin{pmatrix} 1 & -(v+1) \\ 1 & 0 \end{pmatrix}.$$

Let

$$P_v = \begin{pmatrix} 2v\sqrt{4v+3} - 1 & \sqrt{4v+3} - 2 \\ \sqrt{4v+3} - 2 & 4v - 1 \end{pmatrix}.$$

We have

$$P_v^{-1} A P_v = \begin{pmatrix} \dfrac{1}{2} & -\dfrac{\sqrt{4v+3}}{2} \\ \dfrac{\sqrt{4v+3}}{2} & \dfrac{1}{2} \end{pmatrix}.$$

For a suitable function $\gamma(v)$, this matrix can be actually written as

$$(1+v) \begin{pmatrix} \cos\gamma(v) & -\sin\gamma(v) \\ \sin\gamma(v) & \cos\gamma(v). \end{pmatrix}$$

We have so recovered the linear part of the system (3.37). System (3.37) can be actually thought of as the truncation at the third order terms of the Taylor expansion of the righthand side of a more general system. In this sense, the Neimark-Sacher bifurcation is generic. It is characterized by the following theorem (see [2]).

Theorem 3.9 *Let the system*

$$x^+ = f(\mu, x) \tag{3.39}$$

be given, with $x \in \mathbf{R}^2$, $\mu \in \mathbf{R}$, f of class C^4 with respect to both variables in a neighborhood of $(\mu_0, x^*) \in \mathbf{R} \times \Omega$. Assume that:

(i) $f(\mu_0, x_*) = x_*$;
(ii) $(D_x f)(\mu_0, x_*)$ *has a pair of complex (not real) conjugate eigenvalues so that the equation $x = f(\mu, x)$ admits a unique solution $x = \varphi(\mu)$ such that $x_* = \varphi(\mu_0)$, defined in a neighborhood of μ_0 (Lemma 3.1);*
Let $\lambda(\mu), \overline{\lambda(\mu)}$ be the eigenvalues of the matrix $(D_x f)(\mu, \varphi(\mu))$, that we can assume complex conjugate for small variations of μ around μ_0. Assume in addition that:

---

[2] Matrix A is similar, for $v = 0$, to the matrix of the system of Example 1.26.

(iii) $|\lambda(\mu_0)| = 1, (\frac{d}{d\mu}|\lambda(\mu)|)_{\mu=\mu_0} \neq 0$ and $\lambda(\mu_0)^j \neq 1$ for every power $j = 1, 2, 3, 4$.

Then there exists a diffeomorphism (which includes also a change of the parameter) which, after truncation at the 4-th order terms, brings system (3.39) to the form (3.37). If $\alpha(0) \neq 0$, the system undergoes a Neimark-Sacher bifurcation.

3.11 Deterministic Chaos

The determinism is a way of the scientific thinking developed in the framework of the positivism. It was motivated by the extraordinary successes and discoveries of sciences during the seventeenth and eighteenth centuries. Roughly speaking, the principal idea of the determinism is that all the phenomena of the physic world are governed by laws which can be expressed by means of mathematical formulas and equations. The solution of these equations requires in general additional data (initial conditions). Once the equations and the additional data have been exactly determined, one could be able to reconstruct the past history and to predict the future evolution of any phenomenon. The promoters of the determinism were aware that to measure "exactly" a physical quantity is practically impossible. But at that time, this aspect of the problem was probably consider not so crucial. Indeed, if the equations are regular enough and a theorem of continuity with respect to the initial data can be proved, one expects that a small error in the initial data will cause a small error in the final result.[3]

At the end of the nineteenth century, a first event raises a doubt about this belief. During his studies about the three bodies problem, Henri Poincaré realizes that orbits with unexpected properties could be possible in the system governed by the gravitation law. In spite of the regularity of the equations (outsides the singularities), the geometric structure of such orbits is so complicated to make practically impossible the prediction of the future evolution on sufficiently large intervals of time (Poincaré's tangle: [2, 3, 5]). Because of the difficulty of computing explicitly such orbits, Poincaré's intuition did not have immediately a great impact on the scientific community. But around the half of twentieth century, orbits with a similar unpredictability features were discovered and in fact numerically simulated (thank to the new powerful computing tools) in a completely different context by the meteorologist E. Lorenz.

The surprising and apparently paradoxical idea that practical unpredictability and continuity with respect to the initial data can coexist, strongly contributed to the rapidly increasing success and diffusion of what today we call deterministic chaos theory. To this respect, it should be emphasized that complex behaviors of initially near orbits emerge in general only on sufficiently large intervals of time. The meaning of the expression "large intervals of time" depends on the application.

---

[3] At the end of the eighteenth century, the notion of continuity was not yet well formalized.

For weather forecasting, usually the order of magnitude is a few days. In astrophysics, it is probably millions of years.

To formalize the idea of chaotic system, we need to introduce some definitions. Let as usual

$$x^+ = f(x) \tag{3.40}$$

be given, with $f : \Omega \to \Omega$ continuous and $f(\Omega) \subseteq \Omega$.

Definition 3.4 We say that (3.40) has the *dense orbit property* if there exists a point $\hat{x} \in \Omega$ such that the sequence $\{f^{[n]}(\hat{x})\}$ has a dense image in Ω.

Definition 3.5 We say that (3.40) exhibits *sensitive dependence* on initial conditions if there exists a number $\sigma > 0$ such that for each $\hat{x} \in \Omega$ and each neighborhood $\mathcal{U}$ of $\hat{x}$ there exist a point $\hat{y} \in \mathcal{U}$ and an integer n such that $|f^{[n]}(\hat{x}) - f^{[n]}(\hat{y})| > \sigma$.

In the literature, there is not yet a common agreement about the definition of chaotic system. The following one, usually attributed to Devaney [7], is frequently adopted and rather natural.

Definition 3.6 We say that (3.40) is *chaotic* on Ω if:

 (i)   it has the dense orbit property;
 (ii) the set of all its periodic points (of any period) is dense in Ω;
(iii) it exhibits sensitive dependence on initial conditions.

Remark 3.10 It should be noticed that under our assumptions, the three conditions required in Definition 3.6 are not independent. Sensitive dependence is actually a consequence of (i) and (ii) (see [8]). Here, it has been included in accordance with Devaney's definition and since it is often considered as the most important evidence of the chaotic nature of a systems. On the other hand, there are systems which by no means can be said to be "chaotic" and nevertheless, they met the requirements of Definition 3.5 (see next example).   ■

Example 3.12 The linear system $x^+ = 2x$ $(x \in \mathbf{R})$ has a unstable fixed point at the origin. The solutions have the form $x_n = 2^n \hat{x}$. Given any initial state $\hat{x}$ and any $\varepsilon > 0$, let $\hat{y}$ be such that $|\hat{y} - \hat{x}| = \varepsilon$. Then we have $|y_n - x_n| = 2^n |\hat{y} - \hat{x}| = 2^n \varepsilon$. Hence, if $n > \log_2 \frac{1}{\varepsilon}$, we have $|y_n - x_n| > 1$. According to Definition 3.5, this system exhibits sensitive dependence on initial conditions, but of course nobody would say that this system is chaotic.   ■

On the other hand, if (3.40) exhibits sensitive dependence on initial conditions and it admits a fixed point x_*, then x_* is unstable.

Example 3.13 We already know that the map $g(\theta) = 2\theta$ mod 2π on the unit circle (Example 1.18) defines a system with chaotic behavior. Indeed, properties (i) and (ii) of Definition 3.6 have been checked. By applying to the map g the transformation

$y = h(\theta) = -\cos(\theta/2)$ from the unit circle to the interval $[-1, 1]$, one obtains the map $f(y) = 1 - 2y^2 : [-1, 1] \to [-1, 1]$. The map h does not determine a topological equivalence. Nevertheless, being injective, it transforms dense orbits in dense orbits and periodic orbits in periodic orbits (see also [7]). We can conclude that the map f defines a chaotic system. The same conclusion applies to the systems defined by the logistic map (with coefficient $a = 4$) and the tent map (compare with Example 3.4). ∎

References

1. Hartman, P.: Ordinary Differential Equations, 2nd edn. Birkhä user, Boston (1982)
2. Kuznetsov, Y.A.: Elements of Applied Bifurcation Theory. Springer, New York (1995)
3. Guckenheimer, J., Holmes, P.: Ordinary Differential Equations. Springer, New York (1983)
4. Carr, J.: Applications of Centre Manifold Theory. Springer, Berlin (1981)
5. Diacu, F., Holmes, P.: Celestial Encounters. Princeton University Press, Princeton (1996)
6. Palis, J., de Melo, W.: Geometric Theory of Dynamical Systems. Springer, New York (1982)
7. Devaney, R.L.: An Introduction to Chaotic Dynamical Systems. Benjamin-Cummings, Menlo Park (1986)
8. Vesentini, E.: An introduction to topological dynamics in dimension one. Rendiconti del Seminario Matematico dell'Università e del Politecnico di Torino **55**, 303–357 (1997)

Chapter 4
Positive Linear Systems

In many applications it is reasonable to assume that the process is linear. In addition, it frequently happens that the entries of the matrix which defines the system are positive numbers, since they are identified by measuring physical quantities. In this circumstances, the system evolution exhibits special features. The study of this type of systems is the subject of this chapter.

4.1 Positive Matrices

A (row or column) vector $x \in \mathbf{R}^d$ is said to be *positive* if all its components are positive. A square matrix $A \in \mathbf{R}^d \times \mathbf{R}^d$ is said to be *positive* if all its entries are positive.[1] A (row or column) vector $x \in \mathbf{R}^d$ is said to be *nonnegative* if all its components are nonnegative. A square matrix $A \in \mathbf{R}^d \times \mathbf{R}^d$ is said to be *nonnegative* if all its entries are nonnegative.

A linear system of the form

$$x_{n+1} = Ax_n \tag{4.1}$$

is said to be *positive* if A is a positive matrix. It is said to be *nonnegative* if A is nonnegative.

---

[1] Not to be confused with "positive definite"; it is not difficult to find examples of 2×2 matrices which are positive but not positive definite, and vice-versa.

© The Author(s), under exclusive license to Springer Nature Switzerland AG 2022
A. Bacciotti, *Discrete Dynamics*, Mathematical Engineering,
https://doi.org/10.1007/978-3-030-95092-7_4

4.1.1 Perron's Theorem

The distinguishing properties of positive matrices are stated by the following classical theorem, due to O. Perron (published in 1907). For a possible proof, the reader is referred to [1].

Theorem 4.1 (Perron's Theorem) *Let A be a positive matrix. Then, there exists a real eigenvalue r_0 of A such that:*

(i) $r_0 = \rho(A) > 0$;
(i) $m_a(r_0) = 1$;
(i) for each eigenvalue λ of A, if $\lambda \neq r_0$ then $|\lambda| < r_0$;
(i) there exists an eigenvector v_0 of A associated to r_0 such that v_0 is positive.

The eigenvalue r_0 is called the *dominant* eigenvalue. In fact, r_0 is the unique eigenvalue of A which admits a positive eigenvector.

Proposition 4.1 *Let A be a positive matrix, and let λ be any eigenvalue of A. If there exists a positive eigenvector v of A associated to λ then $\lambda = r_0 = \rho(A)$.*

Proof Since A is positive, the matrix A^t is positive as well. Moreover, we know that A and A^t have the same eigenvalues and so the same spectral radius. By Perron's Theorem, there exists a positive vector u_0 such that

$$A^t u_0 = r_0 u_0 \tag{4.2}$$

where as before $r_0 = \rho(A)$. Now assume by contradiction that there exists an eigenvalue $\lambda \neq r_0$ of A with a positive eigenvector w i.e., $Aw = \lambda w$. From (4.2) we get

$$r_0 w^t u_0 = w^t A^t u_0 = (Aw)^t u_0 = \lambda w^t u_0.$$

This yields $(r_0 - \lambda) w^t u_0 = 0$. Since $\lambda \neq r_0$, we get $w^t u_0 = 0$. But this is impossible, since all the components of w and u_0 are positive. ∎

Example 4.1 The eigenvalues of the 3×3 matrix

$$A = \begin{pmatrix} 1 & 1 & 1 \\ 1 & 1 & 1 \\ 1 & 1 & 1 \end{pmatrix}$$

are 3, with $m_a = 1$ (dominant eigenvalue) and 0, with $m_a = m_g = 2$. As a positive eigenvector associated to $\lambda = 3$ we can take $v = (1, 1, 1)$. The eigenspace associated to $\lambda = 0$ is generated, for instance, by $v_1 = (1, 0, -1)$ and $v_2 = (0, 1, -1)$. It is clear that positive linear combinations of v_1 and v_2 cannot exists. ∎

Example 4.2 The eigenvalues of the 4×4 matrix

$$A = \begin{pmatrix} 1 & 1 & 1 & 7 \\ 1 & 1 & 1 & 1 \\ 1 & 7 & 1 & 1 \\ 1 & 1 & 1 & 1 \end{pmatrix}$$

are 6 (dominant eigenvalue), -2 and 0. A positive eigenvalue associated to $\lambda = 6$ is $(2, 1, 2, 1)$. For the eigenvalue $\lambda = 0$ this time we have $m_a = 2$ and $m_g = 1$. In particular, this matrix is not diagonalizable. ∎

4.1.2 Dynamical Consequences

We give now a first look at how the positiveness assumption about A affects the dynamical behavior of the discrete linear system (4.1). We limit ourselves to the case where the eigenvalues of A are real and distinct, but the extension to the general case is absolutely trivial. Let $r_0 = \rho(A), r_1, \ldots, r_{d-1}$ be the eigenvalues of A. Let us agree that the indices have been chosen in such a way that

$$r_0 > |r_1| \geq \cdots \geq |r_{d-1}| \geq 0.$$

Let $v_0, \ldots, v_{d-1}$ be the corresponding linearly independent eigenvectors (v_0 positive). The general solution write:

$$x_n = c_0 r_0^n v_0 + c_1 r_1^n v_1 + \cdots + c_{d-1} r_{d-1}^n v_{d-1}$$
$$= r_0^n \left(c_0 v_0 + c_1 \frac{r_1^n}{r_0^n} v_1 + \cdots + c_{d-1} \frac{r_{d-1}^n}{r_0^n} v_{d-1} \right)$$

for arbitrary constants $c_0, \ldots, c_{n-1}$. Since $|r_i|/r_0 < 1$, this implies

$$x_n = r_0^n (c_0 v_0 + o(1))$$

for $n \to \infty$. We distinguish the following cases.

- $r_0 > 1$. All the solutions for which $c_0 \neq 0$ diverge. The direction of $\{x_n\}$ approaches, for $n \to \infty$, the direction of v_0.
- $r_0 = 1$. All the points on the straight line generated by v_0 (in fact, only these points) are fixed points. All the solutions starting outside this line approach asymptotically the point $c_0 v_0$ (the limit of every solution is finite and depends on the initial state).
- $r_0 < 1$. All the solutions converge to the origin.

It is worthwhile noticing that if $r_0 \leq 1$ the system cannot have periodic solutions (apart from the trivial ones). This is obvious if $r_0 < 1$. As far as the case $r_0 = 1$ is concerned, assume that $\hat{x}$ is a periodic point of period $p \geq 2$. Then $\hat{x}$ is a fixed point of A^p. But actually, A^p is positive, its spectral radius is 1 and its eigenvectors are the same as those of A. Hence the fixed points of A^p are also fixed points of A.

Example 4.3 The eigenvalues of the 2×2 matrix

$$A = \begin{pmatrix} 2 & 1 \\ 1 & 2 \end{pmatrix}$$

are 3 (corresponding eigenvector $(1, 1)$) and 1 (corresponding eigenvector $(1, -1)$). The line generated by $(1, -1)$ is formed by fixed points for the associated system. All of them are unstable.   ∎

Example 4.4 The eigenvalues of the 2×2 matrix

$$A = \begin{pmatrix} 1 & 1 \\ 4 & 1 \end{pmatrix}$$

are 3 (corresponding eigenvector $(1, 2)$) and -1 (corresponding eigenvector $(1, -2)$). The eigenvalues of A^2 are 9 and 1. Hence, the system associated to A has a line formed by periodic points of period 2. All the other solutions diverges. If the initial state is not on the line generated by $(1, 2)$, they have an oscillatory behavior. All the periodic solutions are unstable.   ∎

Example 4.5 The eigenvalues of the 3×3 matrix

$$A = \begin{pmatrix} 1 & 1 & 2 \\ 2 & 1 & 1 \\ 1 & 2 & 1 \end{pmatrix}$$

are 4 and $\frac{-1 \pm i \sqrt{3}}{2}$. The matrix A^3 has the eigenvalue 1 with $m_a = m_g = 2$. Hence, it is diagonalizable. The system associated to A has a two-dimensional subspace formed by periodic solutions of period 3. All the other solutions diverges.   ∎

The assumption that all the entries of the matrix A are strictly positive may be too restrictive for same applications. The next sections deal with possible generalizations of Perron's Theorem, aiming to weaken such assumption.

4.2   Primitive Matrices

A square matrix $A \in \mathbf{R}^d \times \mathbf{R}^d$ is said to be *primitive* if it is nonnegative and there exists an integer $k \geq 1$ such that the matrix A^k is positive. Obviously, every positive matrix is primitive, but the converse is not true (the following example is taken from [1], p. 78).

Example 4.6 For the 3×3 matrix

$$A = \begin{pmatrix} 0 & 1 & 1 \\ 1 & 0 & 0 \\ 1 & 1 & 1 \end{pmatrix}$$

we have

$$A^3 = \begin{pmatrix} 2 & 3 & 3 \\ 2 & 1 & 1 \\ 4 & 4 & 4 \end{pmatrix}$$

∎

Notice that if A is primitive and A^k is positive, then A^m is positive for each integer $m \geq k$. The following proposition may be of help in order to check whether a matrix is primitive, (see again [1], p. 81).

Proposition 4.2 *A nonnegative matrix $A \in \mathbf{R}^d \times \mathbf{R}^d$ is primitive (if and) only if $A^{(d-1)^2+1}$ is positive.*

The following theorem enlightens the interest in primitive matrices.

Theorem 4.2 *All the conclusions (i), (ii), (iii), (iv) of Theorem 4.1 remain valid under the assumption that A is primitive.*

4.3 Nonnegative Matrices in General

Of course, the first natural question is what it is possible to say when the available information is only that A is nonnegative. For a proof of next theorem, see again [1], p. 84.

Theorem 4.3 *Let A be a nonnegative $d \times d$ real matrix. Then, there exists an eigenvalue r_0 of A such that:*

(i') $r_0 = \rho(A) \geq 0$;
(iv') *there exists an eigenvector v_0 of A associated to r_0 such that v_0 is nonnegative.*

Now we give some examples to show that if A is nonnegative but not positive, the eigenvalue $r_0 = \rho(A)$ could have algebraic multiplicity greater than 1. Moreover, we could have eigenvalues $\lambda \neq r_0$ for which $|\lambda| = r_0$.

Example 4.7 The identity matrix in $\mathbf{R}^2$

$$A = \begin{pmatrix} 1 & 0 \\ 0 & 1 \end{pmatrix}$$

possesses a unique eigenvalue 1 with $m_a = m_g = 2$. All the points $x \in \mathbf{R}^2$ are fixed points for the associated dynamical system (4.1). ∎

Example 4.8 The matrix

$$A = \begin{pmatrix} 1 & 1 \\ 0 & 1 \end{pmatrix}$$

possesses the unique eigenvalue 1 with $m_a = 2$, $m_g = 1$. The proper subspace is generated by $(1, 0)$ and there exists no positive eigenvector. The line generated by the eigenvector $(1, 0)$ is formed by fixed points. All of them are unstable. The general solution of the associated system is $x_n = (c_1 + nc_2, c_2)$.   ■

Example 4.9 The matrix

$$A = \begin{pmatrix} 0 & 1 \\ 1 & 0 \end{pmatrix} \tag{4.3}$$

possesses the eigenvalues 1 and -1. Again, we have a line generated by the eigen-vector $(1, 1)$ corresponding to the eigenvalue $\lambda = 1$ formed by fixed points. This time all of them are stable. Indeed, the other points are all periodic of period 2.   ■

Example 4.10 The matrix

$$A = \begin{pmatrix} 0 & 0 & 1 \\ 1 & 0 & 0 \\ 0 & 1 & 0 \end{pmatrix} \tag{4.4}$$

possesses three distinct eigenvalues $(1, \frac{-1 \pm i\sqrt{3}}{2})$ all of modulus 1. The associated system (4.1) has a line of stable fixed points, generated by the eigenvector $(1, 1, 1)$ corresponding to the eigenvalue $\lambda = 1$. All the other points are periodic of period 3.   ■

Example 4.11 The eigenvalues of the matrix

$$A = \begin{pmatrix} 0 & 1 & 1 & 0 \\ 1 & 0 & 0 & 1 \\ 0 & 0 & 0 & 1 \\ 0 & 0 & 1 & 0 \end{pmatrix}$$

are 1 and -1. Both have $m_a = 2$ and $m_g = 1$. The associated system (4.1) has a one-dimensional subspace of fixed points and one-dimensional subspace of periodic points. All of them are unstable.   ■

4.4 Reducible Matrices

Recall that $P \in \mathbf{R}^d \times \mathbf{R}^d$ is said to be a *permutation* matrix if its columns are obtained by any permutation of the columns (or rows) of the identity matrix. If P is a per-mutation matrix and A is any square matrix of the same dimensions, the matrix PA

has the same rows of A written in a different order; the matrix AP has the same columns of A written in a different order. A permutation matrix P is orthogonal. Hence, $P^t P = I$ and $\det(P) = \pm 1$.

Notation In the present and next sections, given a matrix A we denote by A_{ij} a (not necessarily square) submatrix of A. A submatrix is called also a block. A submatrix whose entries are all zero is denoted by $\boxed{0}$.

Definition 4.1 Let $d \geq 2$. A matrix $A \in \mathbf{R}^d \times \mathbf{R}^d$ is called *reducible* if there exists a permutation matrix $P \in \mathbf{R}^d \times \mathbf{R}^d$ such that

$$P^t A P = \begin{pmatrix} A_{11} & A_{12} \\ \boxed{0} & A_{22} \end{pmatrix} \tag{4.5}$$

where the diagonal blocks A_{11}, A_{22} are square submatrices i.e., there exists some integer m, with $1 < m < d$ such that $A_{11} \in \mathbf{R}^m \times \mathbf{R}^m$ and $A_{22} \in \mathbf{R}^{d-m} \times \mathbf{R}^{d-m}$. It is called *irreducible* if it is not reducible.

Clearly, in order to be reducible a matrix must have some zero entry. Therefore, every positive matrix is irreducible. Moreover, it is easy to check that if A is reducible, then every power of A is reducible, as well. Hence, a simple contradiction argument leads to the following conclusion.

Proposition 4.3 *Every primitive matrix is irreducible.*

Matrix (4.3) is an example of a nonnegative and irreducible matrix which is not primitive. In fact, the simple example of matrix (4.3) can be consider a motivation for the main results of this Chapter, exposed in the next section. The following examples and remarks aim to clear up the notions of reducibility/irreducibility.

Example 4.12 The matrix

$$A = \begin{pmatrix} 2 & 0 \\ 1 & 3 \end{pmatrix}$$

is reducible. Indeed,

$$\begin{pmatrix} 0 & 1 \\ 1 & 0 \end{pmatrix} A \begin{pmatrix} 0 & 1 \\ 1 & 0 \end{pmatrix} = \begin{pmatrix} 3 & 1 \\ 0 & 2 \end{pmatrix}.$$

∎

Example 4.13 The matrix

$$A = \begin{pmatrix} 0 & 0 & 0 & 1 \\ 0 & 0 & 1 & 0 \\ 0 & 1 & 0 & 0 \\ 1 & 0 & 0 & 0 \end{pmatrix}$$

is reducible. Indeed, setting

$$P = \begin{pmatrix} 1 & 0 & 0 & 0 \\ 0 & 0 & 0 & 1 \\ 0 & 0 & 1 & 0 \\ 0 & 1 & 0 & 0 \end{pmatrix}$$

we find

$$P^t A P = \begin{pmatrix} 0 & 1 & 0 & 0 \\ 1 & 0 & 0 & 0 \\ 0 & 0 & 0 & 1 \\ 0 & 0 & 1 & 0 \end{pmatrix}$$

∎

Remark 4.1 The term "reducible" comes from the theory of linear algebraic systems. Assume we want to solve the system

$$Ax = b \tag{4.6}$$

where A is a square matrix for which (4.5) holds. It is easily checked that solving (4.6) is equivalent to solve the system

$$P^t A P y = c \tag{4.7}$$

where $y = P^t x$ and $c = P^t b$. The solution of (4.7) can be found by a two-steps cascaded procedure: first one solves the subsystem

$$A_{22} v = q \tag{4.8}$$

and then the subsystem

$$A_{11} u = -A_{12} v + p \tag{4.9}$$

where v, u, q, p are vectors of the appropriate dimensions such that $y = (u, v)$ and $c = (p, q)$.

From the computational point of view, the solution of the pair of reduced order subsystems (4.8) and (4.9) is more convenient than the direct solution of (4.6). ∎

Remark 4.2 Let $e_1, \ldots, e_d$ be the canonical basis of $\mathbf{R}^d$, and let A be a matrix of the form

$$\begin{pmatrix} A_{11} & A_{12} \\ \boxed{0} & A_{22} \end{pmatrix}. \tag{4.10}$$

Let $m \in \{1, \ldots, d-1\}$ denote the dimension of the submatrix A_{11}. Clearly for each index $s \leq m$, the subspace $V_s = \text{span}\{e_1, \ldots, e_s\}$ is algebraically invariant under the action of A i.e., $AV_s \subseteq V_s$.

Vice-versa, if A is an arbitrary matrix such that $AV_s \subseteq V_s$ for each $s \leq m$, then A must have the form (4.10). If P is a permutation matrix, a change of coordinates $x = Py$ determines a reordering of the vectors $e_1, \ldots, e_d$. It follows that the existence of some nontrivial algebraically invariant coordinate subspace (i.e., a subspace of the form $\text{span}\{e_{i_1}, \ldots, e_{i_m}\}$ for same choice of the e_{i_s}'s) characterizes the class of reducible matrices. $\blacksquare$

Remark 4.3 A great help to decide whether a matrix is reducible is provided, at least for not so much high dimensions, by graph theory. To any nonnegative $\mathbf{R}^d \times \mathbf{R}^d$ matrix $A = (a_{ij})$ we associate a graph G with d vertices numbered from 1 to d. The graph is constructed by drawing, for each pair of indices i, j, an oriented arc from vertex i to vertex j if and only if $a_{ij} > 0$. The graph G is called *strictly connected* if for each pair of indices i, j it is possible to reach vertex j from vertex i following a path formed by oriented arcs. A matrix A is irriducible if and only if the associated graph G is strictly connected. Coming back to Example 4.13, it is not difficult to see that vertex 1 of the graph associated to A can be connected to vertex 4, but not to vertices 2 and 3. $\blacksquare$

Another convenient criterion to check whether a matrix is irriducible is provided by the following proposition [1], p. 101.

Proposition 4.4 *Let A be a nonnegative matrix. A is irreducible if and only if $A + A^2 + \cdots + A^d$ is positive.*

The identity (4.5) says that if A is reducible, then it is similar to a block-triangular matrix. However, reducibility is not invariant by more general similarity relations.

Example 4.14 The matrix

$$A = \begin{pmatrix} 2 & 2 \\ 1 & 3 \end{pmatrix}$$

is positive and so irriducible. It is diagonalizable via the matrix

$$Q = \begin{pmatrix} 1 & 2 \\ 1 & -1 \end{pmatrix}$$

and therefore similar to the matrix

$$\begin{pmatrix} 4 & 0 \\ 0 & 1 \end{pmatrix}$$

which is, clearly, reducible. $\blacksquare$

4.5   Frobenius' Theorem

The most valuable result about irriducible nonnegative matrices is the following partial extension of Perron's Theorem, due to F. G. Frobenius (published in 1912). Its proof is far from being trivial and can be carried out in several ways: to this respect, the reader will find some comments in the bibliographical note at the end of this chapter.

Theorem 4.4 (Frobenius' Theorem, first part) *Let a matrix* $A \in \mathbf{R}^d \times \mathbf{R}^d$ *be non-negative and irreducible. Then, the conclusions (i), (ii) and (iv) of Theorem 4.1 hold.*

However, under the assumptions of Frobenius' Theorem matrix A may admit some eigenvalue λ such that $|\lambda| = \rho(A)$ but $\lambda \neq \rho(A)$. This happens for instance with the matrix (4.3).

Definition 4.2 Let $A \in \mathbf{R}^d \times \mathbf{R}^d$. The *index of imprimitivity h* of A is the number of the distinct eigenvalues λ of A such that $|\lambda| = \rho(A)$ $(1 \leq h \leq d)$. A matrix is called *regular* if its imprimitivity index is $h = 1$.

If A is primitive, then by Theorem 4.2 it is regular. The identity matrix is nonnegative and regular, but not primitive. However, if A is nonnegative, regular and irreducible, then it is actually primitive (see [2], p. 290: we warn the reader that in [2] the terminology is different from our). In other words, if we limit ourselves to nonnegative irreducible matrices, the notions of primitive and regular matrix are equivalent.

Theorem 4.5 (Frobenius' Theorem, second part) *Let* $A \in \mathbf{R}^d \times \mathbf{R}^d$ *be nonnegative and irreducible. Then, for all the eigenvalues λ of A such that $|\lambda| = \rho(A)$ one has $m_a = 1$. Moreover, the following statements are equivalent:*

- λ *is an eigenvalue of A such that* $|\lambda| = \rho(A)$;
- λ *is a solution of the equation* $z^h - \rho(A)^h = 0$;
- $\lambda = \rho(A)e^{i\frac{2\pi m}{h}}$ *for some* $m = 0, 1, \ldots, h - 1$;

where h is the index of imprimitivity of A. In addition, the image of any eigenvalue λ of A under a rotation of the complex plane of an angle $2\pi/h$ is still an eigenvalue of A with the same algebraic multiplicity. Finally, if $h > 1$, there exists a permutation matrix P such that

$$P^t A P = \begin{pmatrix} 0 & A_{12} & 0 & \cdots & 0 \\ 0 & 0 & A_{23} & \cdots & 0 \\ \cdots & \cdots & \cdots & \cdots & \cdots \\ 0 & 0 & 0 & \cdots & A_{h-1,h} \\ A_{h,1} & 0 & 0 & \cdots & 0 \end{pmatrix} \qquad (4.11)$$

where the zero diagonal blocks $A_{ii} = \boxed{0}$ *are square (canonic irreducibility form).*

A consequence of the invariance of the spectrum of A under rotations of an angle $2\pi/h$ is that for any eigenvalue λ of A of modulus $|\lambda| < \rho(A)$, the number of eigenvalues of A having the same modulus as λ must be an integer multiple of h.

Example 4.15 Let

$$M = \begin{pmatrix} 0\,0\,1 \\ 1\,0\,0 \\ 0\,1\,0 \end{pmatrix}$$

(the same matrix already considered in Example 4.10) and

$$A = \begin{pmatrix} \boxed{0} & M \\ M & \boxed{0} \end{pmatrix}$$

where the zero blocks are 3×3. By means of the graph criterion, we may easily check that A is irriducible. The eigenvalues of A are $1, -1, (1 \pm i\sqrt{3})/2, (-1 \pm i\sqrt{3})/2$. All have modulus 1. The imprimitivity index of A is 6. ∎

Example 4.16 Let

$$S = \begin{pmatrix} 1\,1\,2 \\ 1\,2\,1 \\ 2\,1\,1 \end{pmatrix}$$

and

$$A = \begin{pmatrix} \boxed{0} & \boxed{0} & S \\ S & \boxed{0} & \boxed{0} \\ \boxed{0} & S & \boxed{0} \end{pmatrix}$$

where the zero blocks are 3×3. Again, the graph criterion shows that A is irriducible. There are three eigenvalues $4, 2(-1 \pm i\sqrt{3})$ of modulus 4, and 6 eigenvalues of modulus 1, placed on the vertices of a hexagon. The imprimitivity index of A is 3. ∎

Next example shows that the rotational symmetric structure of the eigenvalue set can be lost, if the irreducibility assumption is not met.

Example 4.17 Let I be the 2×2 identity matrix and let M be as in Example 4.15. Let finally

$$A = \begin{pmatrix} I & 0 \\ 0 & M \end{pmatrix}.$$

Here, the zero blocks are rectangular, respectively 2×3 and 3×2. We have the eigenvalue 1 with $m_a = m_g = 3$ and, in addition, the eigenvalues $-1, \frac{-1 \pm i\sqrt{3}}{2}$ all of modulus 1.   ∎

4.6 Stochastic Matrices

A nonnegative matrix $W = (w_{ij}) \in \mathbf{R}^d \times \mathbf{R}^d$ is said to be *stochastic* (by rows) if $\sum_{j=1}^{d} w_{ij} = 1$ for each $i = 1, \ldots, d$. It is not difficult to see that if W is stochastic, then every power W^k is stochastic, as well.

Let W be a stochastic matrix and let $\mathbf{1}$ be the column vector of $\mathbf{R}^d$ whose components are all equal to 1. Of course, $W\mathbf{1} = \mathbf{1}$ which means that 1 is an eigenvalue of W and $\mathbf{1}$ is a corresponding eigenvector. On the other hand, if W is nonnegative and it has the eigenvalue 1 with corresponding eigenvector $\mathbf{1}$, then it is stochastic.

In this chapter we already encountered examples of stochastic matrices, or matrices which can be rendered stochastic if multiplied by a scalar factor. For instance, the matrix $W = \frac{1}{3}A$, where A is the matrix of Example 4.1, is stochastic. The interest in the study of stochastic matrices is motivated by the following statement.

Proposition 4.5 *For any stochastic matrix W, one has $\rho(W) = 1$.*

Proof Since 1 is an eigenvalue of W, we have $\rho(W) \geq 1$. On the other hand, from linear algebra we know that $\rho(W) \leq ||W||$ for any induced matrix norm. In particular, W being stochastic, we have

$$\rho(W) \leq ||W||_\infty = \max_i \sum_{j=1}^{d} |a_{ij}| = \max_i \sum_{j=1}^{d} a_{ij} = 1.$$

∎

The previous remarks point out that if W is stochastic, the discrete system

$$x^+ = Wx \tag{4.12}$$

always admits infinitely many fixed points. All of them are critical and fill a subspace of dimension at least 1. The system associated to the matrix of Example 4.10 has indeed a one-dimensional subspace of fixed points. For the system associated to the identity matrix of $\mathbf{R}^d$ all the points of $\mathbf{R}^d$ are fixed points. A stochastic matrix may have also eigenvalues $\lambda \neq 1$ of modulus equal to 1 (see for instance the matrices (4.3) and (4.4)) and, of course, eigenvalues of modulus less than 1. The system associated to a stochastic matrix may have periodic points.

Next we discuss the stability of the fixed points of a system defined by a stochastic matrix. We point out that by virtue of the results of Sect. 2.1.2, and recalling that if λ is an eigenvalue of W then λ^k is an eigenvalue of W^k, the following statements can be extended to periodic points.

Proposition 4.6 *If W is stochastic, all the solutions of (4.12) are bounded.*

Proof The statement follows from the inequality $|x_{n+1}|_\infty = |Wx_n|_\infty \leq ||W||_\infty \cdot |x_n|_\infty = |x_n|_\infty$. ∎

Referring to Theorem 2.12, and recalling that for a linear system the stability of any fixed point is equivalent to stability of the origin, we can state a stability result.

Corollary 4.1 *All the fixed points of (4.12) are stable.*

These results have also an interesting consequence on the geometric structure of a proper basis associated to a stochastic matrix.

Theorem 4.6 *If W is stochastic and λ is an eigenvalue of W with $|\lambda| = 1$, then $m_a(\lambda) = m_g(\lambda)$.*

Proof We give the proof for the case $\lambda = 1$. Assume, by contradiction, that there is a generalized eigenvector u generated by the eigenvector $\mathbf{1}$, that is $(W - I)u = \mathbf{1}$. Let $\{x_n\}$ be the solution of (4.12), for which $x_0 = a\mathbf{1} + bu$ $(b \neq 0)$. Arguing by induction, we readily see that $x_n = (a + nb)\mathbf{1} + bu$. Hence, this solution is unbounded and this is a contradiction to Proposition 4.6. ∎

The following example shows that if W has some eigenvalue λ with $|\lambda| < 1$, it may happens that $m_a(\lambda) > m_g(\lambda)$.

Example 4.18 The matrix

$$W = \begin{pmatrix} 1/2 & 1/4 & 1/4 \\ 0 & 1/2 & 1/2 \\ 0 & 0 & 1 \end{pmatrix}$$

has, as expected, the eigenvalue 1 with $m_a = 1$. In addition, it has the eigenvalue $1/2$ with $m_a = 2$ and $m_g = 1$. This fact does not affect the stability of the fixed points, since the solutions issuing from initial states having a nonzero component along the generalized eigenvector approach the origin (the corresponding eigenvalues being less than 1 in modulus). ∎

Example 4.19 The matrix

$$W = \frac{1}{5} \begin{pmatrix} 1 & 1 & 2 & 1 \\ 1 & 1 & 1 & 2 \\ 1 & 1 & 1 & 2 \\ 1 & 1 & 1 & 2 \end{pmatrix}$$

has the eigenvalue 1 and, in addition, the zero eigenvalue with $m_a = 3$ and $m_g = 2$. Note that here W is positive. ∎

Of course, none of fixed points of system (4.12) defined by a stochastic matrix can be attractive. However, under additional assumptions the attraction set of each fixed point can be completely identified.

Proposition 4.7 *Let W be any stochastic matrix. Assume that W is regular and let $m_a(1) = 1$. Then, the attraction set $\mathcal{A}(x_*)$ of each fixed point x_* is a $(d-1)$-dimensional linear manifold. The subspace parallel to $\mathcal{A}(x_*)$ does not depend on x_* and is generated by the eigenvectors and generalized eigenvectors relative to the eigenvalues $\lambda \neq 1$.*

As a matter of fact, the previous proposition can be recovered as a consequence of other well known and more precise results concerning stochastic matrices (see for instance [1–3]) that we report below without proof.

Theorem 4.7 *Let W be any stochastic matrix. There exists a (unique) matrix L such that $\lim_{n \to \infty} W^n = L$ if and only if W is regular.*

In particular, the limit exists if W is primitive. Notice however that in the previous theorem it is not required that $m_a(1) = 1$ (see for instance the trivial case of the identity matrix).

Theorem 4.8 *Let W be any stochastic matrix, and assume that it is regular. If in addition W is irreducible, then there exists a vector $c \in \mathbf{R}^d$ such that $\lim_{n \to \infty} W^n = \mathbf{1}c^t$.*

Note that $\mathbf{1}c^t$ is a $d \times d$ matrix whose rows are all equal. Under the assumptions of Theorem 4.8, for each solution of the dynamical system (4.12) one has

$$\lim_{n \to \infty} x_n = \lim_{n \to \infty} W^n x_0 = \mathbf{1}c^t x_0 = (c^t x_0)\mathbf{1}.$$

In other words, the limit of $\{x_n\}$ is a vector whose components are equal (and hence, it is a fixed point) and depend on the initial state.

4.7   Bibliographical Note

Frobenius' Theorem, very often quoted as Perron-Frobenius Theorem, has been in the last decades one of the most popular tools of applied mathematics. We limit ourselves to recall some examples.

Compartmental models employed typically in biodynamics are linear discrete dynamical systems defined by matrices whose entries represent concentrations of biological quantities, and are so measured by nonnegative numbers.

The so-called Markov chains, introduced by the Russian mathematician A. A. Markov, are actually linear discrete dynamical systems, defined by stochastic matrices $W = (w_{ij})$: here, w_{ij} represents the probability that a system changes its state from a configuration i to a configuration j.

More recently, Frobenius' Theorem has been one of the key tools of the success of social network dynamics (see for instance the tutorial [4, 5]). It is also related to the so called *PageRank algorithm*, employed by World Wide Web search engines.

As far as its proof is concerned, apart from the original Frobenius paper, it is dutiful quoting the book [3]. However, many other different proofs have been published later and reported in the literature. For a detailed survey, the reader is referred to [6], but a look to the Wikipedia article may be sufficient to get a glimpse. Partial proofs can be found in [1, 2]. See also [7].

References

1. Pullman, N.J.: Matrix Theory and Its Applications. Dekker, New York (1976)
2. Lancaster, P.: Theory of Matrices. Academic Press, New York (1969)
3. Gantmacher, F.R.: Théorie des Matrices, Tome 1. Dunod, Paris (1966)
4. Proskurnikov, A.V., Tempo, R.: A tutorial on modeling and analysis of dynamical social networks. Part I Annu. Rev. Control **43**, 65–79 (2017)
5. Proskurnikov, A.V., Tempo, R.: A tutorial on modeling and analysis of dynamical social networks. Part II Annu. Rev. Control **45**, 166–190 (2018)
6. MacCluer, C.R.: The many proofs and applications of Perron's theorem. SIAM Rev. **42**(3), 487–498 (2000)
7. Shur, A.M.: Detailed Proof of the Perron-Frobenius Theorem. Ural Federal University (2016). Available online

Index

Printed in the United States
by Baker & Taylor Publisher Services